Advances in

VIRUS RESEARCH

VOLUME **75**

Natural and Engineered Resistance to Plant Viruses

Advances in **VIRUS RESEARCH**

VOLUME **75**

Natural and Engineered Resistance to Plant Viruses

Edited by

GAD LOEBENSTEIN

Agricultural Research Organization
Bet Dagan, Israel

JOHN P. CARR

Department of Plant Sciences
University of Cambridge, U.K.

ELSEVIER

AMSTERDAM • BOSTON • HEIDELBERG • LONDON
NEW YORK • OXFORD • PARIS • SAN DIEGO
SAN FRANCISCO • SINGAPORE • SYDNEY • TOKYO
Academic Press is an imprint of Elsevier

Academic Press is an imprint of Elsevier

32 Jamestown Road, London, NW17BY, UK
Radarweg 29, PO Box 211, 1000 AE Amsterdam, The Netherlands
30 Corporate Drive, Suite 400, Burlington, MA 01803, USA
525 B Street, Suite 1900, San Diego, CA 92101-4495, USA

First edition 2009

Library of Congress Cataloging-in-Publication Data

A catalog record for this book is available from the Library of Congress

British Library Cataloguing-in-Publication Data

A catalogue record for this book is available from the British Library

ISBN: 978-0-12-381397-8
ISSN: 0065-3527

Printed and bound in USA
09 10 11 12 10 9 8 7 6 5 4 3 2 1

CONTENTS

PREFACE

Since the very earliest developments in agriculture, and probably even before then, diseases affecting crop plants have posed an ever-present, yet ever changing, threat to human survival. The Bible, for example, explicitly mentions blights, blasts, and mildew diseases of wheat. Not surprisingly, people sought to understand and mitigate the effects of disease on crop productivity, and many earlier cultures have sought divine aid in the fight against crop disease. The Romans, according to some historians, celebrated the festival of Robigalia: an attempt to mollify Robigus, the god thought to protect crops from disease, and his less benign sister Robiga (or Robigo), a primary goddess of Roman farmers, known as the spirit of mildews and rusts. However, even during this period there were attempts to understand plant diseases through the application of reason: an approach exemplified in the writings of Theophrastus (372–287 BC), who theorized about the nature of the diseases of cereals and other plants. Meanwhile, over many centuries farmers all over the world practiced domestication of plants from wild populations and selected the best and hardiest plants grown under agricultural conditions, thereby incidentally breeding plants resistant to disease.

In the modern world the deployment of crops possessing genetically based resistance is generally considered the best and most economical approach for disease control. This is especially true for protection against viruses because, so far at least, no chemicals are available that could provide the same degree of protection in the field against these pathogens, as fungicides do against fungi and oomycetes. The transfer by breeding of naturally occurring resistance genes from wild plants or land races to cultivated lines is still an ongoing process, and has been supplemented with other methods such as mutation, polyploidy breeding, and the generation of haploids. Genetic resistance against virus diseases can be surprisingly durable. A good example is that of cucumbers bred for resistance to *Cucumber mosaic virus*. This resistance, which depends on several genes, was found to be stable for many decades against different strains of this virus.

Even though the majority of plants are resistant to most viruses (the phenomenon of non-host or basal resistance), when viruses *are* able to infect a crop plant, obtaining durable resistance by breeding is not always possible. In certain cases, new virus strains overcome the resistance and once again may cause severe crop losses. In addition, for some crops and viruses, no suitable sources of resistance can be identified among the wild relatives of a crop plant. Hence the need for greater understanding of natural resistance, and for the insights its study can provide for the development of novel crop protection approaches.

In the last few years, much has been learned concerning the mechanisms underlying several natural resistance mechanisms including *inter alia* RNA silencing, induced resistance, and resistance conferred by recessive and dominant genes, which will be discussed in this and the following volume of the *Advances*. In addition, research over the last two decades has made it possible to move resistance–conferring gene sequences between plants from different botanical genera, or into plants from other organisms, and even from the viruses themselves (*pathogen-derived resistance*). This work opened a new vista for plant virus control, and if combined with engineering for insect resistance could potentially provide protection not only against the viruses themselves, but also against their vectors. The work on pathogen-derived resistance also led directly to the discovery of a natural resistance and gene regulation mechanism, RNA silencing, that has ramifications throughout the whole of biomedicine. Nevertheless, these technologies face technical and sociological challenges, which are also addressed in these volumes.

In all parts of the world, but especially among the developing nations, agriculture faces the looming problems of emerging virus diseases, population growth, and ecological change. We hope that the articles in this volume and the following one will inform and stimulate research on natural and engineered resistance, and thereby contribute to the development of new approaches to disease control and the creation of new resistant varieties that are desperately needed.

We want to thank Professor Karl Maramorosch for inviting us to edit these thematic volumes; all of our contributors who have prepared comprehensive, stimulating, and thought-provoking reviews; the technical staff of the *Advances*, and specifically Mr. Ezhilvijayan Balakrishnan and Ms. Narmada Thangavelu from Chennai, India.

October 2009

Gad Loebenstein
John P. Carr
Editors

CHAPTER 1

Mechanisms of Recognition in Dominant *R* Gene Mediated Resistance

P. Moffett[*,†]

Contents

Abstract

One branch of plant innate immunity is mediated through what is traditionally known as race-specific or gene-for-gene resistance wherein the outcome of an attempted infection is determined by the genotypes of both the host and the pathogen. Dominant plant disease resistance (*R*) genes confer resistance to a variety of biotrophic pathogens, including viruses, encoding corresponding dominant avirulence (*Avr*) genes. *R* genes are among the most highly variable plant genes known, both within and between populations. Plant genomes encode hundreds of *R* genes that code for NB-LRR

[*] Boyce Thompson Institute for Plant Research, Tower Road, Ithaca NY 14853, USA
[†] Département de Biologie, Université de Sherbrooke, 2500 Boulevard de l'Université, Sherbrooke, Québec, Canada J1K 2R1

Advances in Virus Research, Volume 75
ISSN: 0065-3527, DOI: 10.1016/S0065-3527(09)07501-0

proteins, so named because they posses nucleotide-binding (NB) and leucine-rich repeat (LRR) domains. Many matching pairs of NB-LRR and Avr proteins have been identified as well as cellular proteins that mediate R/Avr interactions, and the molecular analysis of these interactions have led to the formulation of models of how products of *R* genes recognize pathogens. Data from multiple NB-LRR systems indicate that the LRR domains of NB-LRR proteins determine recognition specificity. However, recent evidence suggests that NB-LRR proteins have co-opted cellular recognition co-factors that mediate interactions between Avr proteins and the N-terminal domains of NB-LRR proteins.

I. DOMINANT GENETIC RESISTANCE TO PATHOGENS

A. Plant innate immunity

Pathogens exert significant constraints on host fitness. These pressures have consequently applied strong selection on the evolution of host genomes such that most hosts devote significant portions of their genomes to encoding defenses against pathogens. Like all multi-cellular organisms, plants possess an innate immune system: a system of defense against pathogens based on germline-encoded components. In many cases, the innate immune system functions to recognize pathogen-associated molecular patterns (PAMPs) via receptor-like proteins known as pathogen recognition receptors (PRRs). PRRs are often well conserved in structure and function and recognize PAMPs that are also well conserved and associated with broad classes of pathogens, such as bacterial flagellin, fungal chitin and various components of bacterial cell walls (Nicaise *et al.*, 2009). Responses induced by PRRs, often referred to as PAMP-induced immunity (PTI), are generally "low-impact" and are sufficient to confer resistance to most pathogens (Chisholm *et al.*, 2006). However, host-adapted pathogens are able to overcome PTI mechanisms through the deployment of so-called "effector" proteins. These proteins are delivered to the host cytoplasm via the various secretion mechanisms of bacteria and eukaryotic pathogens and for the most part it appears that the main function of these proteins is to interfere with PTI signaling (Chisholm *et al.*, 2006; Guo *et al.*, 2009). In turn, plants have also evolved several classes of receptor-like proteins that are more specific in their recognition spectra. These plant proteins recognize specific pathogen-associated proteins and induce a much more drastic suite of "high impact" defense responses, often culminating in the induction of a type of programmed cell death known at the hypersensitive response (HR). Since many of the pathogen proteins that induce these responses are effector proteins, this is often referred to as effector-triggered immunity (ETI) (Chisholm *et al.*, 2006). The plant proteins that induce ETI are highly expanded in number, and highly variable both within and between species. The inherent variability of these proteins manifests as differences in resistance or susceptibility to pathogens within a plant species and has led to the genetic identification of the loci responsible for this variability.

B. Disease resistance genes

The contribution of single major gene to resistance to pathogens in plants was initially noted by Biffen upon the popularization of Mendel's laws, who reported on the presence of recessive sources of pathogen resistance in wheat (Biffen, 1905). Recessive resistance is common in plants and its molecular basis has been reviewed elsewhere (Iyer-Pascuzzi and McCouch, 2007; Robaglia and Caranta, 2006). In many cases, recessive resistance results from the inability of the pathogen to infect the host due to a lack of compatibility between pathogen-encoded factors and the host proteins they need to interact with to establish an infection. This is particularly well-defined for recessive plant resistance to viruses (Robaglia and Caranta, 2006) and can be viewed in essence, not as a recognition of the pathogen by the plant, but as a lack of recognition of the plant by the pathogen.

In a series of studies, Flor documented the existence of dominant disease resistance (*R*) genes in different cultivars of flax which conferred resistance to specific strains of flax rust (Flor, 1971). Resistance was also dependent on the presence of genes in the pathogen that rendered the pathogen avirulent, but only on those host genotypes possessing a corresponding *R* gene (Fig. 1). Like *R* genes, these avirulence (*Avr*) genes were also dominant, suggesting an active recognition process on the part of the plant that responds to specific pathogen-associated molecules. Thus, the result of an attempted infection is dependent on both the genotype of the host and that of the pathogen and as such, this form of resistance is known as gene-for-gene resistance. Gene-for-gene relationships have since been shown to exist between hosts and many other pathogens and pests, including insects, nematodes, fungi, oomycetes, bacteria and viruses (Martin *et al.*, 2003). A large number of dominant *R* genes have been cloned (Sacco and Moffett, 2009) which encode for a relatively small number of receptor-like protein classes (Fig. 2). A number of *R* genes encode proteins with transmembrane and extracellular leucine-rich repeat (LRR) domains. To date, these proteins, known as LRR receptor-like proteins (LRR-RLPs), have only been shown to confer resistance to fungal pathogens, recognizing proteins secreted from the pathogen into the host apoplast. In addition, two rice *R* genes conferring resistance to a bacterial pathogen (*Xa21* and *Xa3*/*Xa26*)

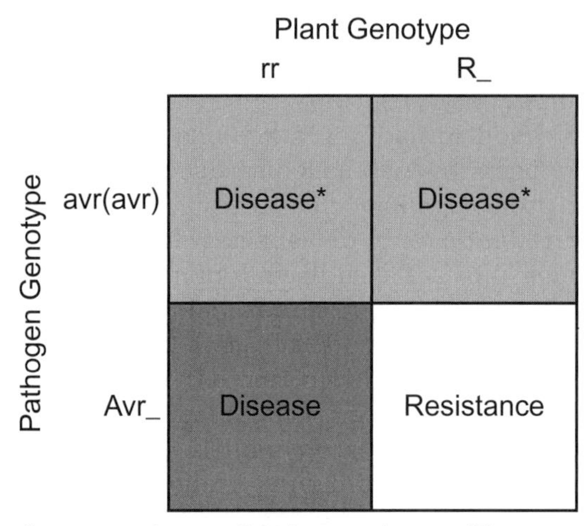

FIGURE 1 Gene-for-gene resistance. Dominant resistance (*R*) genes confer resistance to specific pathogens. This resistance is dependent on whether the pathogen possesses a matching avirulence (*Avr*) gene or not (*avr*). Asterisks and lighter gray shading indicate that in many cases, mutation or deletion of the *Avr* gene results in a fitness cost for the pathogen such that although it gains virulence on resistant hosts, it suffers a fitness penalty on susceptible hosts.

encode proteins of similar structure, but with the addition of an intracellular receptor-like kinase domain (LRR-RLKs; Fig. 2). However, since AvrAx21 is thought to be a sulfated peptide rather than a protein, it has been proposed that Xa21 may be more appropriately considered as a PRR similar to several PRRs which also encode LRR-RLKs (Lee *et al.*, 2006; Nicaise *et al.*, 2009).

The majority of cloned *R* genes encode for NB-LRR proteins; so named for the presence of a conserved nucleotide-binding (NB) and a C-terminal LRR domain (Fig. 2). NB-LRR-encoding genes are rapidly diversifying and make up one of the largest and most variable gene families found in plants (Clark *et al.*, 2007), with 149 members in *Arabidopsis*, 317 in poplar, 54 in papaya, 400–500 in *Medicago*, 233 in grape and 480 identified in rice (Ameline-Torregrosa *et al.*, 2008; Kohler *et al.*, 2008; Meyers *et al.*, 2003; Porter *et al.*, 2009; Velasco *et al.*, 2007; Zhou *et al.*, 2004). To date, over seventy *R* genes encoding NB-LRR proteins with known resistance specificities have been cloned, which confer resistance to the gamut of plant pathogens, including insects, nematodes, oomycetes, fungi, bacteria and viruses (Sacco and Moffett, 2009). Importantly, there do not appear to be any characteristics that define the NB-LRR proteins that recognize different pathogens. That is, the NB-LRR proteins that recognize bacteria, for example, cannot be distinguished from those that recognize viruses. Indeed, very similar

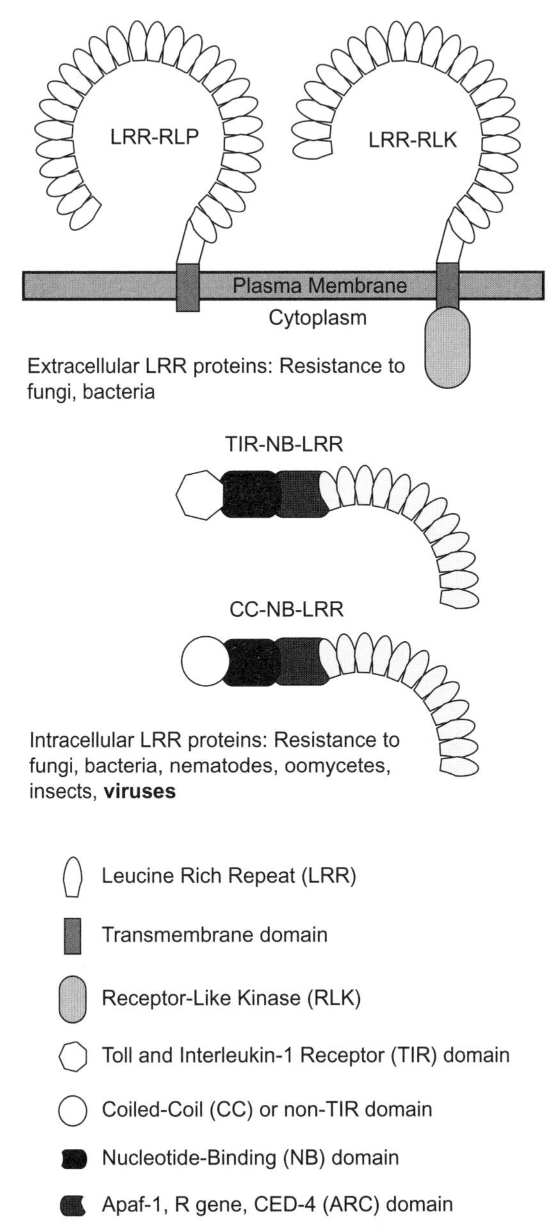

FIGURE 2 Schematic diagram of the proteins encoded by disease resistance genes along with a list of the types of pathogens recognized by these proteins. These include proteins with extracellular moieties (top) or intra-cellular proteins (middle). The identity of individual protein domains are indicated at the bottom.

NB-LRR proteins are capable of recognizing completely different pathogens, as in the cases of closely related Rx and Gpa2 proteins of potato which recognize a virus and a nematode, respectively; and the allelic *Arabidopsis* NB-LRR proteins RPP8, HRT and RCY1 which confer resistance to an oomycete and two different viruses (Bendahmane *et al.*, 2000; Cooley *et al.*, 2000; McDowell *et al.*, 1998; Takahashi *et al.*, 2002; van der Vossen *et al.*, 2000). Likewise, NB-LRR proteins encoded at orthologous genomic loci in the *Solanaceae* appear to have retained a certain level of sequence homology but have evolved to recognize different pathogen types (Grube *et al.*, 2000; Mazourek *et al.*, 2009). Furthermore, the resistance response initiated is not specific to the class of pathogen detected. Delivery of bacterial or oomycete Avr proteins from a viral vector to plants encoding the corresponding NB-LRR protein results in resistance to the recombinant virus or bacteria (Rentel *et al.*, 2008; Tobias *et al.*, 1999). These observations suggest that once activated, NB-LRR proteins induce responses that are effective against all pathogen types, allowing for flexibility with respect to the types of pathogens that a given NB-LRR protein can evolve to recognize. In terms of understanding function however, this suggests that information gleaned from different experimental systems can be used to elucidate what underlying molecular mechanisms govern the function of most NB-LRR proteins.

II. NB-LRR PROTEIN STRUCTURE

A. The domains of NB-LRR proteins

1. The NB-ARC domain

The NB and LRR domains are the most commonly identified domains present in NB-LRR proteins. Between these two domains however is a region of homology known as the ARC (Apaf-1, R proteins, CED-4) domain (van der Biezen and Jones, 1998a). The ARC domain can be further subdivided into the ARC1 and ARC2 domains, two functionally distinct structural entities (Albrecht and Takken, 2006; Rairdan and Moffett, 2006) (Fig. 2). The NB and ARC domains, often referred to collectively as the NB-ARC or NBS (nucleotide-binding site), comprise a functional nucleotide-binding pocket capable of binding and hydrolyzing ATP (Tameling *et al.*, 2002). The NB-ARC region contains a number of motifs highly conserved throughout the members of the STAND (signal transducing ATPases with numerous domains) class of proteins (Rairdan and Moffett, 2007). These include the P-loop, kinase 2, and kinase 3/RNBS-B motifs of the NB domain, the GxP motif of the ARC1 domain as well as the MHDV motif of the ARC2 all of which are thought to contribute to the nucleotide binding and/or hydrolysis (Albrecht and

Takken, 2006; McHale *et al.*, 2006; Tameling *et al.*, 2006). The NB-ARC also contains a number of well-conserved motifs whose functions are unknown but which may be involved in initiating downstream signaling, as over-expression of the NB domain of the Rx protein has been shown to be sufficient to induce defense responses (Rairdan and Moffett, 2007; Rairdan *et al.*, 2008). In addition, characteristic differences between these motifs have allowed for the identification of two main classes of NB-LRR proteins (Cannon *et al.*, 2002; Meyers *et al.*, 1999). These differences also correlate with the type of domain found at the N termini of these two classes.

2. N-terminal domains

Type members of the first class of NB-LRR protein possess a Toll and Interleukin-1 Receptor (TIR) homology domain, a protein–protein interaction domain associated with innate immune systems in multiple phyla (Staal and Dixelius, 2007). Members of the TIR-NB-LRR protein family belong to an ancient lineage with members present in angiosperms, gymnosperms and possibly bryophytes, although this class appears to have been lost in monocots (Akita and Valkonen, 2002; Bai *et al.*, 2002; Liu and Ekramoddoullah, 2003; Meyers *et al.*, 1999). Although the TIR domain appears to be well conserved among family members, the second class of NB-LRR proteins appears to have undergone significantly more diversification within the N terminus as well as elsewhere throughout the protein (Cannon *et al.*, 2002). Since many proteins of this second class appear to encode coiled-coil (CC) motifs within their N termini, these proteins are often referred to as CC-NB-LRRs, or simply as non-TIR NB-LRRs. However, many "CC" domains do not conform to coiled-coil prediction programs and there are few motifs within this domain that are broadly conserved with the exception of the EDVID motif which is present in most characterized CC domains (Mazourek *et al.*, 2009; Rairdan *et al.*, 2008). In addition, the CC domain is often coupled with an additional domain such as the Solanceaous domain (SD) of proteins such as Prf and Mi-1 (Mucyn *et al.*, 2006), or a predicted BED DNA binding domain as in Xa1 (Yoshimura *et al.*, 1998). In addition, a large family of poplar proteins has been identified wherein the CC domain is replaced by a BED domain and many NB-LRR proteins of both the TIR and CC class (as identified by their conserved NB-ARC motifs) lack any domain N-terminal to the NB domain (Kohler *et al.*, 2008). Thus, variability in the N-terminal domains of CC-NB-LRR proteins appears to constitute a significant source of variation in this class of proteins.

The N-terminal domains of NB-LRR proteins have traditionally been thought to act as a signaling domain due to similarities in domain structure between plant NB-LRR proteins and certain animal proteins involved in apoptosis and innate immunity (Rairdan and Moffett, 2007). Indeed, over-expression of several TIR domains has been demonstrated to induce defense signaling (Frost *et al.*, 2004; Swiderski *et al.*, 2009). However, no such function has been reported for CC domains and many recent reports indicate a role for both CC and TIR domains in recognition specificity (see below).

3. The LRR domain

The C-terminal domain of NB-LRR proteins is made up of individual repeats with the consensus LxxLxxLxxLxLxx(N/C/T)x(x)LxxIPxx, although the primary structure and number of individual repeats vary greatly between family members (Jones and Jones, 1997). The LRR domain appears to be under diversifying selection, particularly at residues predicted to be solvent-exposed suggesting that this domain may be involved in determining recognition specificity (Botella *et al.*, 1998; McDowell *et al.*, 1998; Meyers *et al.*, 1998; Mondragon-Palomino *et al.*, 2002; Noel *et al.*, 1999). Consistent with this idea, domain swapping experiments between highly similar NB-LRR proteins with different recognition specificities have shown that recognition specificity maps to the LRR domain (Ellis *et al.*, 2007a; Rairdan and Moffett, 2006; Shen *et al.*, 2003; Zhou *et al.*, 2006). The LRR domain of the rice Pi-Ta protein has been reported to interact with its cognate Avr protein in yeast and *in vitro* (Jia *et al.*, 2000). Furthermore, the allelicly encoded flax L5, L6 and L7 proteins bind to specific versions of the AvrL567 proteins in yeast and this discrimination is determined by their LRR domains (Dodds *et al.*, 2006; Ellis *et al.*, 2007b).

B. Interactions between domains

A key discovery in the elucidation of NB-LRR protein function was the observation that the different domains of these proteins undergo both physical and functional interactions. The LRR domains of Rx, Bs2, HRT, RPS5 and N have all been shown to interact with their cognate N-terminal halves (Ade *et al.*, 2007; Leister *et al.*, 2005; Moffett *et al.*, 2002; Rairdan and Moffett, 2006; Ueda *et al.*, 2006). Although the ARC1 domain is sufficient for this physical interaction, the LRR also undergoes a functional interaction (and possibly a physical interaction) with the ARC2 domain (Rairdan and Moffett, 2006). Domain swapping between closely related NB-LRR paralogues often results in auto-activated molecules; proteins that induce defense responses in the absence of any Avr protein (Howles *et al.*, 2005; Hwang *et al.*, 2000; Rairdan and Moffett, 2006; Sun *et al.*, 2001). In the case of Rx/Gpa2 swaps, this auto-activity is conditioned by inappropriate pairings of the LRR and ARC2 domains, leading us to suggest that these two domains undergo a "perfect fit" that retains the protein in an inactive conformation until perturbed by some recognition-related event (Rairdan and Moffett, 2006).

The CC domain of Rx also undergoes an intra-molecular interaction with the NB-LRR fragment of the protein (Moffett *et al.*, 2002). Although it is not clear what part of the rest of the protein mediates this interaction, a functional nucleotide-binding pocket and the interaction between the LRR and ARC1 domains are prerequisites for this interaction to occur (Rairdan *et al.*, 2008). The interdependency of these intra-molecular interactions suggests that upon translation, NB-LRR proteins undergo a stepwise folding process wherein both the intra-molecular interactions and nucleotide binding are required in order to adopt a conformation that is competent for activation.

III. MODELS OF NB-LRR RECOGNITION

A. The guard hypothesis

Given the diversity of LRR domains, together with the known role of LRRs as protein–protein interaction domains, this originally led to the idea that NB-LRRs might undergo a receptor–ligand interaction with their cognate Avrs. However, such interactions were not easily demonstrated and this lack of success was compatible with the formulation of the guard hypothesis, which suggests that NB-LRR proteins monitor, or guard, host proteins (Dangl and Jones, 2001; Van der Biezen and Jones, 1998b). In this model, rather than surveying for the mere presence of the pathogen, the plant instead perceives alterations to host target molecules mediated by pathogen effector proteins. As such, these targets, or "guardees," become co-factors in recognition. In keeping with the guard model, there are a number of well-defined systems in which recognition by an NB-LRR protein depends on the interaction between an Avr protein and a cellular co-factor. These include a number of effector proteins from *Pseudomonas syringae* such as: AvrB and AvrRpm1which interact with the *Arabidopsis* RIN4 protein, which causes the activation the NB-LRR protein RPM1 (Mackey *et al.*, 2002); AvrRpt2 cleaves RIN4, which causes the activation of the NB-LRR protein RPS2 (Axtell and Staskawicz, 2003; Mackey *et al.*, 2003); AvrPphB cleaves the *Arabidopsis* protein PBS1, resulting in the activation of the NB-LRR protein RPS5 (Shao *et al.*, 2003); AvrPto and AvrPtoB interact with the tomato protein Pto, resulting in the activation of the NB-LRR protein Prf (Pedley and Martin, 2003). One major assumption of the model is that all Avr proteins function to promote pathogen virulence, presumably by inactivating host targets. Indeed, the guard hypothesis can be seen as particularly enticing in the case of bacterial and eukaryotic pathogens as all Avr proteins from these pathogens are presumed to act as effector proteins and interfere with PTI mechanisms (Chisholm *et al.*, 2006; Guo *et al.*, 2009). However, the targeting of the putative guardees listed above does not appear to be important for virulence. The guard hypothesis is also difficult to reconcile with cases were a given pathogen effector protein overcomes resistance but retains virulence activity (Gassmann *et al.*, 2000; Harrison, 2002).

B. The decoy model

Although the guard hypothesis is attractive in providing a mechanism to account for indirect recognition by NB-LRR proteins, recognition specificity does not always correlate with virulence activity. The decoy model has been proposed to accommodate the guard hypothesis in light of these observations (van der Hoorn and Kamoun, 2008). The decoy model is based on the assumption that guardees will be subject to alternating selective pressures depending on the presence or absence of their guarding NB-LRR protein. In the presence of the NB-LRR protein, the guardee would be optimized for Avr interaction, and hence detection. In the absence of the NB-LRR protein, the guardee would be under pressure to evade interaction with pathogen effectors in order to reduce pathogen virulence. Such conditions have been suggested to favor the evolution of molecular decoys, which interact with pathogen Avr proteins and facilitate R protein recognition, but which do not represent true virulence targets (van der Hoorn and Kamoun, 2008; Zhou and Chai, 2008; Zipfel and Rathjen, 2008). That is, decoy proteins can be neutral in terms of pathogen virulence but if they interact with the same effector molecules as the actual virulence targets this will allow them to act as molecular sensor of pathogen virulence activity.

Support for the decoy model is emerging from cases originally proposed to exemplify the guard model. The tomato Ser/Thr kinase Pto was originally thought to be a *P. syringae* virulence target, with the NB-LRR protein Prf acting as a guard, monitoring the action of *P. syringae* AvrPto and AvrPtoB against Pto (Van der Biezen and Jones, 1998b). However, the virulence targets of AvrPto and AvrPtoB now appear to be the kinase domains of the receptor-like kinases CERK1, BAK1, EFR1 and FLS2 (Gimenez-Ibanez *et al.*, 2009; Göhre *et al.*, 2008; Shan *et al.*, 2008; Xiang *et al.*, 2008), which mediate PTI responses, implying that Pto is being used as a kind of bait by Prf to interact with effector proteins that normally target other kinases.

C. Recognition of viral Avrs by NB-LRR proteins

1. ETI from a viral point of view

The concept of ETI is based on the fact that NB-LRR proteins recognize effector proteins from bacterial or eukaryotic pathogens (Chisholm *et al.*, 2006). However, NB-LRR proteins are intra-cellular receptor-like proteins and thus effector proteins are essentially the only proteins to which NB-LRR proteins are exposed from these pathogens. As such, NB-LRR proteins that recognize viral Avr proteins (Table 1) provide an important contrast and potential exceptions that must be applied to any rules pertaining to NB-LRR protein function. Like other pathogens, viruses are subject to PTI mechanisms in the form of RNA silencing which targets their highly structured and/or double-stranded RNA (Ding and Voinnet, 2007). Like other pathogens, viruses have evolved proteins to inhibit this PTI mechanism, known as viral suppressors of RNA silencing (VSRs), which interfere with various components of the RNA silencing machinery (Diaz-Pendon and Ding, 2008). In turn, plant NB-LRR proteins recognize virus proteins and induce defense responses similar to those induced by NB-LRR proteins that recognize other pathogens (Fig. 3). Importantly however, many of the viral proteins recognized by NB-LRR proteins are not VSRs and may not necessarily be analogous to the effector proteins of other pathogens.

Since viruses generally do not have the same potential for genetic redundancy as other pathogen types, resistance-breaking versions of Avr proteins must retain at least some function or the virus becomes less fit (Fig. 1). All viral proteins, including those recognized by NB-LRR proteins, are required for virulence in the broadest sense of the word, which explains, in part, why many viral *R* genes are highly durable (Garcia-Arenal and McDonald, 2003). However, models wherein recognition is linked to interference with PTI mechanisms are not easily applied to viral Avr proteins with structural (coat proteins and movement proteins) or enzymatic (replicases) roles in virus accumulation. Viral proteins are multi-functional by nature and many interfere with the anti-viral PTI mechanism of RNA silencing (Diaz-Pendon and Ding, 2008). For instance, the *Turnip crinkle virus* (TCV) CP is the Avr determinant for HRT and acts as a potent viral suppressor of RNA silencing (VSR) (Cooley *et al.*, 2000; Qu *et al.*, 2003; Thomas *et al.*, 2003). However, this is not a general correlation as none of the other Avrs listed in Table 1 are known to possess VSR activity. The TCV CP also promotes virulence by interfering with the NAC transcription factor TIP, and recognition of the TCV CP by HRT correlates with its ability to bind TIP (Ren *et al.*, 2000, 2005). However, TIP is not required for HRT-mediated resistance to TCV (Jeong *et al.*, 2008). Thus, although viral Avr proteins clearly have virulence function (Schoelz, 2006), it is not yet clear how this function relates to their detection by NB-LRR proteins.

2. Outcomes of recognition by anti-viral *R* genes

Recognition of viruses by NB-LRR proteins is generally assayed by the outcome of an infection which can result in a range of responses (Moffett and Klessig, 2008) [see also Chapter 3 in this volume and Carr *et al.* (in press)].

TABLE 1 Characterized NB-LRR proteins conferring resistance to viruses

Protein	NB-LRR class	Plant	Virus	Avr	References
HRT[¶]	CC	*Arabidopsis*	*Turnip crinkle virus*	CP	Cooley *et al.* (2000)
RCY1[¶]	CC	*Arabidopsis*	*Cucumber mosaic virus*	CP	Takahashi *et al.* (2002)
Rx	CC	Potato	*Potato virus X* and other potexviruses	CP	Baures *et al.* (2008), Bendahmane *et al.* (1999)
Rx2	CC	Potato	*Potato virus X*	CP	Bendahmane *et al.* (2000), Querci *et al.* (1995)
Sw-5	CC	Tomato	Tospoviruses	ND	Brommonschenkel *et al.* (2000)
Tm-2, Tm-2^2	CC	Tomato	Tobamoviruses	l-30K MP	Lanfermeijer *et al.* (2003), Lanfermeijer *et al.* (2005), Weber *et al.* (1993)
L^1, L^2, L^3, L^4[†]	CC	Pepper	Tobamoviruses	CP	Tomita *et al.* (2008)
Rsv1[†]	CC	Soybean	*Soybean mosaic virus*	P3	Hajimorad *et al.* (2005), Hayes *et al.* (2004)
Ctv[†]	CC	Trifoliate orange	*Citrus tristeza virus*	ND	Rai (2006), Yang *et al.* (2003)
I[†]	TIR	Bean	*Bean common mosaic virus*	ND	Vallejos *et al.* (2006)
N	TIR	*Nicotiana glutinosa*	Tobamoviruses	Helicase: P50 subunit	Erickson *et al.* (1999), Whitham *et al.* (1994)
Y-1	TIR	Potato	*Potato virus Y*	ND	Vidal *et al.* (2002)

CP, coat protein; MP, movement protein; ND, not determined.
[¶]Alleles of the same locus in *Arabidopsis*.
[†]The *L* allelic series, *Rsv1*, *Ctv* and *I* genes map to clusters of NB-LRR encoding genes. The individual genes conferring resistance have yet to be definitively assigned.

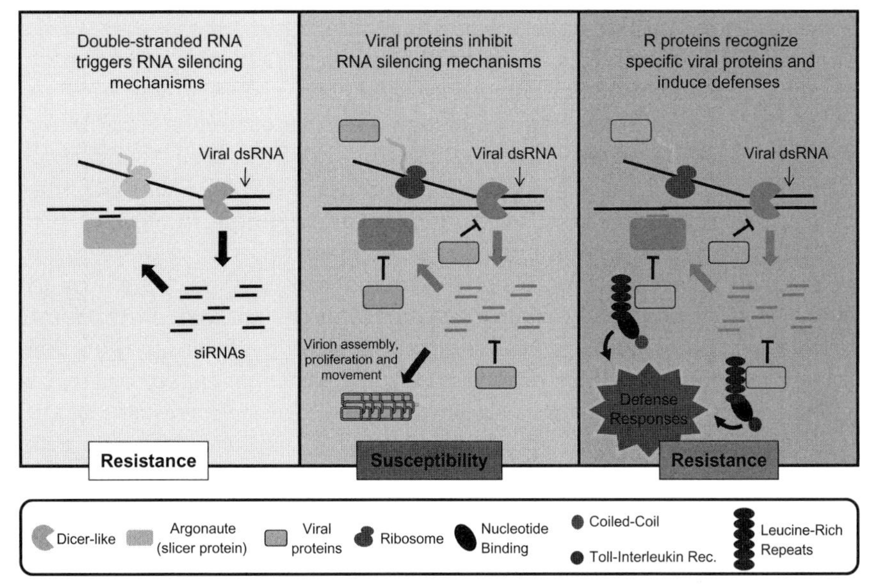

FIGURE 3 PTI and ETI from a viral point of view, adapted from Chisholm *et al.* (2006). (Left) PAMP-triggered immunity (PTI). Recognition of highly structured or double-stranded RNA (dsRNA) by Dicer-like endoribonucleases leads to the cleavage of viral RNAs into small interfering RNAs (siRNAs) of 21–24 nucleotides. These siRNAs are incorporated into slicing complexes of which Argonaute proteins are the main enzymatic component. Argonaute proteins employ siRNAs as guides to target homologous RNAs derived from the virus. These combined activities, known as RNA silencing, result in the degradation of viral RNAs, thus preventing viral protein production, and protect plants from most viruses (see also Chapter 2 in this volume). (Middle) Plant viruses target RNA silencing at various steps, allowing for the accumulation and translation of viral RNAs and subsequent virus spread. (Right) "Effector-triggered" immunity. Plant NB-LRR proteins recognize specific viral proteins and induce a suite of defenses that limit virus accumulation and often result in programmed cell death. (See Page 1 in Color Section at the back of the book.)

At one end of the spectrum is complete susceptibility, and at the other is what is known as extreme resistance (ER), wherein no viral accumulation is detectable and no other response is visible at the site of infection. Very often however, viruses elicit an HR-type resistance, which is accompanied by some viral accumulation and movement outside the initially infected cell, followed by the induction of an HR. Below this level of response is what is often referred to as trailing necrosis or systemic HR (SHR) wherein the defense response induced by an *R* gene is not sufficient to stop viral spread but eventually does induce HR responses in infected tissue. Indeed, in a number of cases the necrotic symptoms induced by viral infection appears to be due to an inefficient responses mediated by NB-LRR proteins (Fujisaki *et al.*, 2004; Kaneko *et al.*, 2004; Lee *et al.*, 1996). The different outcomes associated with anti-viral defense responses are conditioned by a number of factors, including the efficiency of recognition, signaling and viral movement. For example, the *HRT* gene confers HR-type resistance to TCV in the *Arabidopsis* ecotype Di-17, but SHR when introgressed into ecotype Col-0 (Kachroo *et al.*, 2000). This appears to be due to at least one modifying locus which alters levels of the defense signaling molecule salicylic acid (SA), and over-expression of HRT results in ER-type resistance (Chandra-Shekara *et al.*, 2004). This suggests that the different responses are due to differences in the ability of HRT to signal in the different backgrounds rather than differences in ability to recognize the TCV CP. The responses of other *R* genes, such as the *I* gene of bean and *RCY1* of *Arabidopsis*, are also affected by genetic background (Collmer *et al.*, 2000; Takahashi *et al.*, 2001). In addition the *Rx* and *Rx2* genes of potato, both of which confer ER-type resistance in potato to *Potato virus X* (PVX), behave differently when transferred as transgenes to *Nicotiana* spp.: strains of PVX which are entirely resistance-breaking in potato induce SHR in *Rx* transgenic *N. benthamiana*, whereas *Rx2* confers HR-type resistance in transgenic tobacco and *N. benthamiana* (Baures *et al.*, 2008; Bendahmane *et al.*, 2000; Farnham and Baulcombe, 2006; Sacco *et al.*, 2007). Since Rx confers resistance to wild-type PVX equally well in *Nicotiana* as in potato, this suggests that either Rx and Rx2 behave differently in terms of recognition, or that different versions of the PVX CP behave differently, between these species.

3. Recognition of viral Avr proteins

Differences in infection outcome can also occur due to differences between inoculated viruses rather than the context of the NB-LRR protein. For example, the *Sw-5* gene of tomato induces either an ER, HR or no response to different tospoviruses, suggesting that the Sw-5 protein recognizes the different versions of the tospovirus Avr protein to greater or lesser extents, with other factors such as its ability to signal, being equal (Brommonschenkel *et al.*, 2000). The former case illustrates an important observation regarding recognition of viruses by NB-LRR proteins. Plant *R* genes are often cloned due to their ability to confer resistance to a particular virus, but in fact are often able to recognize multiple viruses within the same family which may possess varying degrees of homology in the primary structure of their Avr proteins. For example the *N* gene, originating from *N. glutinosa*, recognizes all known tobamoviruses with a single geographically restricted exception, *Pepper obuda virus* (ObPV) (Padgett and Beachy, 1993; Whitham *et al.*, 1994). Likewise the allelic tomato genes *Tm-2* and *Tm-2*^{2}, introgressed into tomato from different accessions of *Lycopersicum peruvianum*, control nearly all tobamoviruses. However, despite the fact that these two genes encode CC-NB-LRR proteins which differ by only four amino acids, *Tm-2*^{2} has proven to be more durable than *Tm-2* (Lanfermeijer *et al.*, 2003, 2005). This seems to be due to the fact that strains of *Tomato mosaic virus* (ToMV) that overcome *Tm-2* or *Tm-2*^{2} have mutations in different regions of the viral movement protein (MP); some mutations can overcome *Tm-2*, but not *Tm-2*^{2}, and vice versa. However, those mutations that overcome *Tm-2*^{2} gene appear to have a greater impact on virus fitness as evidenced by reduced viral load in plants infected with these strains, which likely explains the greater durability of this gene (Weber *et al.*, 2004).

The interaction between *Tm-2* or *Tm-2*^{2} and the ToMV MP is typical of many NB-LRR experimental systems in that loss of avirulence function is associated with a compromise of the virulence function of the affected protein. However, it is of interest to consider cases where a lack of recognition of a given Avr protein is not associated with a compromise of function. This may be best illustrated with *R* genes that confer resistance to multiple viruses. For example, the *Rx* gene confers resistance to PVX, for which potato is a natural host (Bendahmane *et al.*, 1999). When transferred to *N. benthamiana*, the Rx protein is capable of recognizing the CP of several other potexviruses that infect this host, despite a low degree ($\sim$40%) of identity between these proteins (Baures *et al.*, 2008). However, *Rx* does not confer resistance to the *Carlavirus Poplar mosaic virus* (PopMV), and in theory, the PopMV CP ($\sim$40% identity to the PVX CP) can be considered a non-recognized version of this Avr. It is tempting to speculate that the PopMV CP might not be recognized by Rx because it has different properties or cellular interacting partners (guardees or decoys) than the PVX CP. However, an important study has shown that Rx has the potential to recognize the PopMV CP. Artificial evolution, through random mutagenesis of the Rx LRR domain, resulted in the identification of several mutants that were capable of conferring resistance to a resistance-breaking strain of PVX and combining these mutations resulted in a version of Rx, Rx(M3), capable of recognizing the PopMV CP (Farnham and Baulcombe, 2006). Thus, recognition is not likely due to differences in function or interactions with different cellular co-factors by the different coat proteins but appears to be driven by the Rx LRR domain. Importantly, Rx(M3) retained its ability to recognize wild-type and resistance-breaking strains of PVX indicating that the recognition capacity of Rx had been expanded rather than qualitatively altered.

An interesting natural parallel to Rx(M3) can be seen with the pepper *L* locus whose alleles, L^1, L^2, L^3 and L^4, originate from different pepper species and code for highly similar CC-NB-LRR proteins (Tomita *et al.*, 2008) (K. Kobayashi, pers. comm.). Like the case of Rx and Rx(M3), resistance conferred by the L proteins is characterized by a series of proteins with an increasing breadth of recognition. Tobamoviruses can be classified by their ability to infect plants with different *L* alleles. Those viruses controlled by L^1, L^2, L^3 and L^4 are referred to as being of the P_0 pathotype. Viruses of the P_1 pathotype are not controlled by L^1 whereas $P_{1,2}$, $P_{1,2,3}$ and $P_{1,2,3,4}$ pathotypes overcome the L^2, L^3 and L^4 genes, respectively. The Avr determinant for all four L proteins is the tobamoviral CP (Berzal-Herranz *et al.*, 1995; Gilardi *et al.*, 2004). The coat proteins of the tobamoviruses recognized by L proteins show greater homology (60–75%) than the potexvirus coat proteins recognized by Rx and, within a given virus species, breakage of resistance can be achieved by single, but often double, point mutations (Antignus *et al.*, 2008; Genda *et al.*, 2007). Strains of *Pepper mild mottle virus* (PepMV) that have mutated to overcome *L3* and *L4* may have sustained a minor loss of virulence (Antignus *et al.*, 2008; Genda *et al.*, 2007; Sakamoto *et al.*, 2008). However, it may be more informative to compare PepMV with *Paprika mild mottle virus* (PMMoV) and ToMV, which are controlled by L^2, or L^2 and L^1, respectively. The coat proteins of all three of these viruses are wild type, yet are differentially recognized by the different versions of L, similar to the case of PVX and PopMV recognition by Rx and Rx(M3). Together these examples suggest that recognition is based on some conserved structural of functional aspect of the coat proteins of these viruses but that the different NB-LRR protein variants differ in their "affinity" for this feature.

Unlike the detection of antigens or small molecular signatures in other immune systems, it appears that recognition by NB-LRR proteins does require some minimal secondary structural and/or functional aspect of Avr proteins. For example, recognition of the PVX by Rx does not require a fully functional CP capable of multimerization and formation of virions (Moffett *et al.*, 2002). However, for recognition Rx does require a minimal CP fragment of approximately ninety residues which encompass a predicted four α-helix bundle within the CP (Baures *et al.*, 2008; Nemykh *et al.*, 2008). This suggests that Rx may recognize a conserved structure rather than a particular sequence, which would allow it to recognize multiple potexvirus coat proteins which presumably have very similar structures despite low primary sequence identities. At the same time, Rx may recognize some functional aspect of this CP fragment, although it is not clear what this function might be. On the other hand, structure and function may be difficult to separate as in the case of the NIa protease of *Potato virus Y* (PVY) which functions as the avirulence determinant for the potato *Ry* gene, which is likely (but not demonstrated) to encode a typical NB-LRR protein (Mestre *et al.*, 2000). Mutational analysis of NIa showed that it is not possible to identify mutants that lack protease activity while retaining the ability to elicit Ry, although it is possible to generate NIa mutants that retain protease activity, but do not elicit Ry (Mestre *et al.*, 2003). Likewise, in the case of the tombusvirus VSR protein P19, which induces an HR on tobacco where it is very likely recognized by an endogenous NB-LRR protein, loss of VSR virulence function correlates with loss of HR induction in most, but not all cases (Chu *et al.*, 2000; Hsieh *et al.*, 2009; Scholthof *et al.*, 1995). Thus, it would appear that although NB-LRR proteins may not necessarily recognize viral Avr proteins as a consequence of their function, it may be that only functional proteins present the molecular attributes recognized by NB-LRR proteins.

Understanding how pathogens overcome recognition is a question that is highly pertinent to deploying effective broad range and durable resistance. Clearly, this is often achieved by mutations that result in a loss, or functional compromise, of the Avr determinant. However, as outlined above, several anti-viral *R* genes provide examples where perfectly functional Avr variants are not recognized by one NB-LRR protein, but clearly possess the ability to be recognized by another. Indeed, it has been suggested some time ago that a loss of avirulence can at times be a gain of function (van Loon, 1987). For example, the *N′* gene, which has not been cloned, but which may encode an NB-LRR protein, confers HR-type resistance to most tobamoviruses but does not recognize the U1 strain of TMV. However, certain mutations in the TMV U1 CP can result in a gain of recognition by *N′*; despite the fact that these mutations are not in the hydrophobic pocket of TMV thought to be recognized by *N′* as well as by the L1 protein and a putative R protein present in eggplant (Dardick *et al.*, 1999). Instead, recognized and non-recognized versions of the TMV CP differ in the stoichiometry of their low-order aggregates (dimers, trimers, tetramers), which may in turn alter which surfaces of the TMV CP are exposed for potential recognition (Culver, 2002). This suggests that TMV U1 has evolved to mask some feature common to most or all tobamoviruses that is inherently recognizable by N′, L1 and potentially other NB-LRR proteins. A similar situation can be seen with the NB-LRR protein N. Mutagenesis of ObPV resulted in the identification of a mutation in the p50 protein that allowed recognition by N (Padgett and Beachy, 1993), again suggesting that ObPV has gained the ability to mask recognition rather than lost a virulence function. However, the fact that loss of avirulence often requires multiple mutations in viral Avrs suggests that such masking may require compensatory mutation elsewhere within the Avr protein in order to retain virulence function (Garcia-Arenal and McDonald, 2003; McDonald and Linde, 2002; Parlevliet, 2002).

D. The bait and switch model

As predicted by both the guard and decoy models, a number of proteins have been identified that mediate the recognition of Avr proteins by

NB-LRR proteins. These recognition co-factors include; the RIN4 protein, which interacts with the Avr proteins AvrB, AvrRpm1 and AvrRpt2 as well as the NB-LRR proteins RPM1 and RPS2 (Axtell and Staskawicz, 2003; Mackey *et al.*, 2002, 2003); the Pto protein, which interacts with the Avr proteins AvrPto and AvrPtoB as well as the NB-LRR protein Prf (Kim *et al.*, 2002; Mucyn *et al.*, 2006, 2009); the PBS1 protein, which interacts with the Avr protein AvrPphB as well as the NB-LRR protein RPS5 (Ade *et al.*, 2007); the NRIP1 protein, which interacts with the Avr protein p50 as well as the NB-LRR protein N (Caplan *et al.*, 2008b). This also seems to be the case for the RanGAP2 protein, which interacts with the NB-LRR proteins Rx and Gpa2 and appears to play a role in the recognition of their respective Avr proteins (Sacco *et al.*, 2007, 2009; Tameling and Baulcombe, 2007). Surprisingly however, despite the documented role of the LRR domain in determining recognition specificity (see above), none of these recognition co-factors interact with the LRR domain. Instead, in all cases (excluding RPS2 where it has not been reported), these recognition co-factors interact with the NB-LRR N-terminal domains; the CC domains of RPM1, RPS5, Rx, and Gpa2; the SD domain of Prf, and the TIR domain of N.

To account for the apparent contradiction in the functionally defined role of the LRR domain and the physical associations of these recognition co-factors we have proposed the "bait and switch" model (Collier and Moffett, 2009). This model posits that, in addition to the LRR domain, the N termini of NB-LRR proteins play a role in recognition, and proteins bound to the latter domain function as molecular baits that mediate a primary interaction with Avr proteins (Fig. 4). In theory, these baits could correspond to either guardee or decoy proteins. For example, the NRIP1 protein interacts with the TIR domain of N and is required for N function (Caplan *et al.*, 2008b). NRIP1 interacts with the Avr determinant of N, the p50 subunit of TMV and mediates an apparent complex between the three proteins (Caplan *et al.*, 2008a). Importantly however, NRIP1 also interacts with a version of p50 with a mutation, derived from the resistance-breaking tobamovirus ObPV (p50-Ob), that abrogates its ability to elicit N-mediated responses (Caplan *et al.*, 2008b). Thus, NRIP1 provides a potential mechanism to explain how N interacts with p50, but does not account for recognition specificity. At the same time, it has been reported that NB-ARC-LRR fragment of N is able to interact with wild-type p50, but not p50-Ob, in yeast two-hybrid assays (Ueda *et al.*, 2006). This is reminiscent of the situation seen with the different NB-LRR protein products of the flax L alleles, which interact with their cognate Avr determinants in yeast, with the LRR domain determining interaction specificity (Dodds *et al.*, 2006; Ellis *et al.*, 2007a). To date, most apparent direct interactions between Avr and NB-LRR proteins have been shown in yeast or *in vitro* suggesting that LRR domains (or LRR/ARC

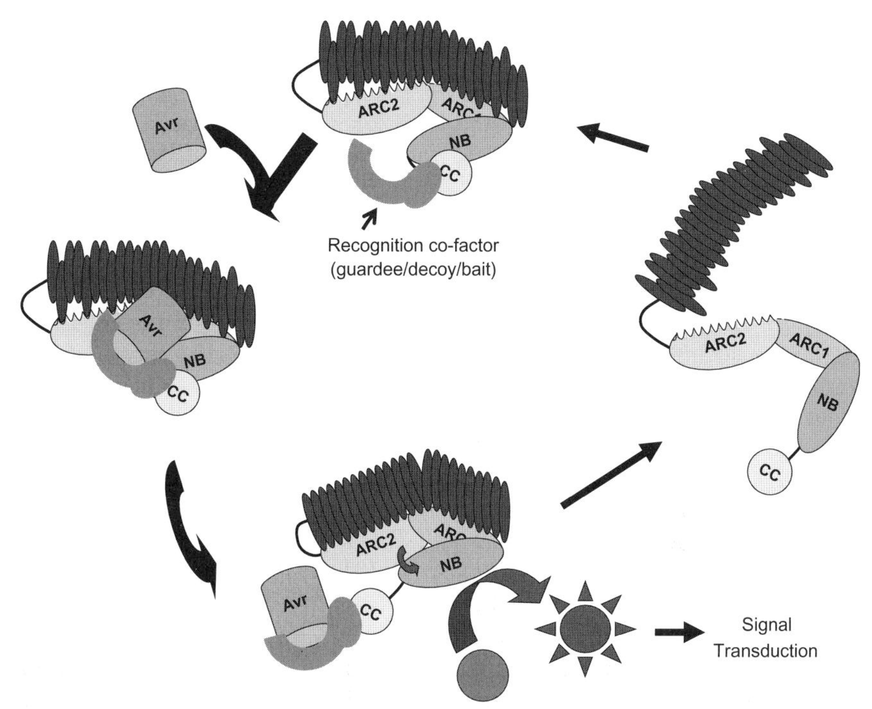

FIGURE 4 The bait and switch model of NB-LRR protein function. Upon translation, NB-LRR proteins are folded in a stepwise fashion into an auto-inhibited state that is primed only once all the individual intra-molecular interactions have occurred, some of which likely require nucleotide binding (top). In many cases, the setting of the "trap" also requires the interaction with a bait protein (blue) with the N-terminal domain (CC, SD, BED or TIR domains) of the NB-LRR protein. In this state the "switch" is in the "off" position. Due to inter-domain interactions involving the N-terminal domain, this may bring the bait protein into close proximity with other domains of the NB-LRR protein. The pathogen Avr protein is brought into the NB-LRR system via the bait protein, either through interaction with the bait and this interaction serves to facilitate interaction between the Avr and the LRR domain (left). Alternatively, the Avr protein may physically alter the bait protein leading to a loss of auto-inhibition. Either of these events lead to perturbation and subsequent conformational changes to the LRR/ARC2 interface causing the switch to be flipped to the molecular switch to the "on" position (bottom). This change is relayed through the ARC2 into an alteration in nucleotide binding status within the nucleotide-binding pocket, which is thought to cause further conformational changes within the nucleotide-binding pocket allow signaling motif(s) within the NB domain (green star) to associate with and activate downstream signaling components. Concomitant with, or subsequent to signaling, intra-molecular interactions within the NB-LRR protein are likely dissociated and reset, allowing the protein to undergo repeated rounds of recognition and signaling (right). (See Page 2 in Color Section at the back of the book.)

"complexes") may have the potential to interact with Avr proteins, but only under certain experimental conditions. However, *in vivo* this interaction may need to be facilitated by an initial interaction with a bait protein which interacts with the N terminus. In the case of N and p50, such a scenario could explain why NRIP1 might interact with both p50 and 50-Ob, but only the former would interact with and activate N. Such a scenario might result in Avr proteins that continue to interact with bait proteins via some relatively conserved structural or functional feature, but that are able to mask subsequent recognition by the LRR domain. This might account for cases where loss of avirulence is due to a gain of function differences in Avr proteins whereas mutations that cause both loss of avirulence and virulence function might be more likely to have altered the ability to interact with the bait protein.

The fact that NRIP1 binds both "recognized" and "non-recognized" versions of the same Avr protein raises the possibility that bait proteins might "present" multiple Avr proteins to their cognate NB-LRR proteins. This might allow the same bait protein to interact with, and mediate recognition by, multiple similar NB-LRR proteins. For example, the Rx and Gpa2 proteins recognize very different proteins, but both require the putative bait protein RanGAP2, which interacts with the nearly identical Rx and Gpa2 CC domains (Sacco *et al.*, 2007, 2009; Tameling and Baulcombe, 2007). Artificial tethering of RanGAP2 to the RBP-1 protein (the Avr for Gpa2) induces a much greater Gpa2-dependent response, but does not induce any Rx-mediated response (Sacco *et al.*, 2009). This can be interpreted to mean that RanGAP2 functions as a bait to mediate an initial interaction with RBP-1, but that ultimately, it is the LRR domain that determines which Avr proteins will cause the NB-LRR protein to be activated.

A variation of the bait and switch model relates to what might be thought of as the "mousetrap model" of NB-LRR function. That is, the intra-molecular interactions of NB-LRR proteins hold the protein in an inactive, but hair-trigger state by that is primed to be sprung. Like the mechanisms of a mousetrap however, springing of the trap can occur by any event that disrupts this delicate balance, including, but not exclusive to, alteration of the bait. Indeed, auto-activating mutations have been found throughout all the different domains of NB-LRR proteins (Takken *et al.*, 2006), suggesting that auto-inhibition can be disrupted in many ways. Furthermore, in several cases, bait proteins appear to act as integral subunits of their cognate NB-LRR partners. These include Pto and RIN4, which are required for accumulation of Prf and RPM1, respectively but appear to retain their NB-LRR partners inactive until some recognition event perturbs this inhibitory function (Balmuth and Rathjen, 2007; Belkhadir *et al.*, 2004; Mucyn *et al.*, 2009). Likewise, the activation of RPS2 and RPS5, upon elimination of their cognate bait proteins, RIN4 and PBS1 by the proteolytic activities of AvrRpt2 and AvrPhB, respectively suggests that these baits play a role in auto-inhibition. This loss of auto-inhibition is thought to then cause alterations in the nucleotide-binding pocket, followed by changes in protein–protein interactions with downstream signaling molecules (Fig. 4). As such, Avr proteins may activate NB-LRR proteins by perturbing either the auto-inhibitory interface between the ARC and LRR domains and/or an auto-inhibitory interaction at the N terminus involving a bait protein. A combination of these mechanisms would allow for activation by specific interactions with the LRR, facilitated by bait proteins, or simply by the direct interference with, or elimination of, the bait protein. To date, most cases that resemble the latter situation come from the recognition of bacterial effectors whereas the former situation might more accurately relate to those situations that have been characterized for virus Avr recognition. However, it is important to note that relatively few well-characterized examples exist of either situation and thus more examples will be needed before generalizations can be drawn. Nonetheless, the bait and switch model provides a conceptual framework to explain apparently contradictory results from different experimental systems, as well to design future studies aimed at understanding the underlying mechanisms that govern NB-LRR recognition, function and evolution. For example, the bait and switch model would predict that in cases such as Rx and Rx(M3), or the L variants, a common bait protein may recognize a very general structure common to a class of coat proteins, but that the individual LRRs may have varying degrees of affinity for different versions thereof. Alternatively, new recognition specificities might arise as NB-LRR proteins change their specificity for different bait proteins which in turn might present a different set of Avr proteins.

IV. PERSPECTIVES

The realization that recognition is mediated in many, or most, cases by recognition co-factor/bait proteins adds a further level of complexity to our understanding of this phenomenon. The existence of bait proteins means that these must be taken into consideration in any attempt to understand how the LRR determines recognition specificity and *vice versa*. Furthermore, any attempts to engineer new recognition specificity must take both bait and LRR into consideration and will require a better understanding of how these two components cooperate to recognize Avr proteins. Likewise, incompatibilities with bait proteins may explain why NB-LRR proteins often fail to function properly, or show altered recognition specificities when transferred between different species

(Alcázar *et al.*, 2009; Bomblies and Weigel, 2007; Ernst *et al.*, 2002; Goggin *et al.*, 2006).

The bait and switch model also raises a number of new questions regarding NB-LRR proteins, particularly with respect to the nature of their associated bait proteins (Collier and Moffett, 2009). Although several putative bait proteins have been identified, it is unclear what, if any, structural or functional commonalities predispose them for this function. Thus it is of great interest to identify additional examples of bait proteins in order to determine what properties have resulted in their being co-opted by the NB-LRR protein surveillance system. Have bait properties been selected because they are frequently targeted by virulence factors as proposed by the guard hypothesis, or because they look like virulence targets as proposed by the decoy model? Alternatively, if some Avr proteins have no obvious role in inhibiting cellular processes, have bait proteins simply been co-opted due to a general propensity to bind to foreign proteins? Presumably, all NB-LRR proteins have evolved from a common progenitor. Within groups of NB-LRR proteins with very similar N termini, it would seem more likely that the bait protein would remain constant with the ARC/LRR interface evolving for its ability to be perturbed by different Avr proteins that interact with the bait. However, with diversification, NB-LRR proteins have clearly evolved to bind different bait proteins. As such, it is of interest to determine how, upon their expansion and diversification, these proteins have evolved to interact with new bait proteins and to determine whether there might be a common set of bait proteins common to all plants. That is, is there a subset of NB-LRR proteins that interact with RIN4, another with Pto-like proteins, another with RanGAP proteins etc., or are bait proteins as diverse as their cognate NB-LRR partners?

Despite a great deal of recent insights into the mechanisms by which NB-LRR proteins recognize pathogens in general, and viruses in particular, there are still many gaps in our knowledge of how these proteins function and a great deal of room for further study. NB-LRR proteins that recognize viruses make up a significant percentage of all NB-LRR proteins and offer a number of distinct experimental systems that will continue to provide insights into how this pathogen surveillance system functions and how it has evolved. In turn, these insights may lead to better strategies for control of diseases caused by viruses and other plant pathogens.

ACKNOWLEDGMENTS

I am grateful to current and former laboratory members for stimulating discussions on NB-LRR protein function. The comments and communication of unpublished results of Kappei Kobayashi are gratefully acknowledged. Work on NB-LRR proteins in the Moffett laboratory at the Boyce Thompson Institute is supported by funding from the National Science Foundation (IOS-0744652).

REFERENCES

Ade, J., DeYoung, B. J., Golstein, C., and Innes, R. W. (2007). Indirect activation of a plant nucleotide binding site-leucine-rich repeat protein by a bacterial protease. *Proc. Natl. Acad. Sci. U. S. A.* **104:**2531–2536.

Akita, M., and Valkonen, J. P. (2002). A novel gene family in moss (*Physcomitrella patens*) shows sequence homology and a phylogenetic relationship with the TIR-NBS class of plant disease resistance genes. *J. Mol. Evol.* **55:**595–605.

Albrecht, M., and Takken, F. L. (2006). Update on the domain architectures of NLRs and R proteins. *Biochem. Biophys. Res. Commun.* **339:**459–462.

Alcázar, R., García, A. V., Parker, J. E., and Reymond, M. (2009). Incremental steps toward incompatibility revealed by *Arabidopsis* epistatic interactions molulating salicylic acid pathway activation. *Proc. Natl. Acad. Sci. U. S. A.* **106:**334–339.

Ameline-Torregrosa, C., Wang, B. B., O'Bleness, M. S., Deshpande, S., Zhu, H., Roe, B., Young, N. D., and Cannon, S. B. (2008). Identification and characterization of nucleotide-binding site-leucine-rich repeat genes in the model plant *Medicago truncatula*. *Plant Physiol.* **146:**5–21.

Antignus, Y., Lachman, O., Pearlsman, M., Maslenin, L., and Rosner, A. (2008). A new pathotype of *Pepper mild mottle virus* (PMMoV) overcomes the *L4* resistance genotype of pepper cultivars. *Plant Dis.* **92:**1033–1037.

Axtell, M. J., and Staskawicz, B. J. (2003). Initiation of RPS2-specified disease resistance in *Arabidopsis* is coupled to the AvrRpt2-directed elimination of RIN4. *Cell* **112:**369–377.

Bai, J., Pennill, L. A., Ning, J., Lee, S. W., Ramalingam, J., Webb, C. A., Zhao, B., Sun, Q., Nelson, J. C., Leach, J. E., and Hulbert, S. H. (2002). Diversity in nucleotide binding site-leucine-rich repeat genes in cereals. *Genome Res.* **12:**1871–1884.

Balmuth, A., and Rathjen, J. P. (2007). Genetic and molecular requirements for function of the Pto/Prf effector recognition complex in tomato and *Nicotiana benthamiana*. *Plant J.* **51:**978–990.

Baures, I., Candresse, T., Leveau, A., Bendahmane, A., and Sturbois, B. (2008). The Rx gene confers resistance to a range of potexviruses in transgenic *Nicotiana* plants. *Mol. Plant-Microbe Interact.* **21:**1154–1164.

Belkhadir, Y., Nimchuk, Z., Hubert, D. A., Mackey, D., and Dangl, J. L. (2004). Arabidopsis RIN4 negatively regulates disease resistance mediated by RPS2 and RPM1 downstream or independent of the NDR1 signal modulator and is not required for the virulence functions of bacterial type III effectors AvrRpt2 or AvrRpm1. *Plant Cell* **16:**2822–2835.

Bendahmane, A., Kanyuka, K., and Baulcombe, D. C. (1999). The Rx gene from potato controls separate virus resistance and cell death responses. *Plant Cell* **11:**781–792.

Bendahmane, A., Querci, M., Kanyuka, K., and Baulcombe, D. C. (2000). Agrobacterium transient expression system as a tool for the isolation of disease resistance genes: Application to the Rx2 locus in potato. *Plant J.* **21:**73–81.

Berzal-Herranz, A., de la Cruz, A., Tenllado, F., Diaz-Ruiz, J. R., Lopez, L., Sanz, A. I., Vaquero, C., Serra, M. T., and Garcia-Luque, I. (1995). The Capsicum L3 gene-mediated resistance against the tobamoviruses is elicited by the coat protein. *Virology* **209:**498–505.

Biffen, R. H. (1905). Mendel's laws of inheritance and wheat breeding. *J. Agric. Sci.* **1:**4–48.

Bomblies, K., and Weigel, D. (2007). Hybrid necrosis: Autoimmunity as a potential gene-flow barrier in plant species. *Nat. Rev. Genet.* **8:**382–393.

Botella, M. A., Parker, J. E., Frost, L. N., Bittner-Eddy, P. D., Beynon, J. L., Daniels, M. J., Holub, E. B., and Jones, J. D. G. (1998). Three genes of the *ArabidopsisRPP1* complex resistance locus recognize distinct *Paronospora parasitica* avirulence determinants. *The Plant Cell* **10:**1847–1860.

Brommonschenkel, S. H., Frary, A., and Tanksley, S. D. (2000). The broad-spectrum tospovirus resistance gene Sw-5 of tomato is a homolog of the root-knot nematode resistance gene Mi. *Mol. Plant-Microbe Interact.* **13:**1130–1138.

Cannon, S. B., Zhu, H., Baumgarten, A. M., Spangler, R., May, G., Cook, D. R., and Young, N. D. (2002). Diversity, distribution, and ancient taxonomic relationships within the TIR and non-TIR NBS-LRR resistance gene subfamilies. *J. Mol. Evol.* **54:**548–562.

Caplan, J., Padmanabhan, M., and Dinesh-Kumar, S. P. (2008a). Plant NB-LRR immune receptors: From recognition to transcriptional reprogramming. *Cell Host Microbe* **3:**126–135.

Caplan, J. L., Mamillapalli, P., Burch-Smith, T. M., Czymmek, K., and Dinesh-Kumar, S. P. (2008b). Chloroplastic protein NRIP1 mediates innate immune receptor recognition of a viral effector. *Cell* **132:**449–462.

Carr, J. P., Lewsey, M. G., and Palukaitis, P. (in press). Signaling in induced resistance. *Adv. Virus Res.*

Chandra-Shekara, A. C., Navarre, D., Kachroo, A., Kang, H. G., Klessig, D., and Kachroo, P. (2004). Signaling requirements and role of salicylic acid in HRT- and rrt-mediated resistance to turnip crinkle virus in *Arabidopsis*. *Plant J.* **40:**647–659.

Chisholm, S. T., Coaker, G., Day, B., and Staskawicz, B. J. (2006). Host-microbe interactions: Shaping the evolution of the plant immune response. *Cell* **124:**803–814.

Chu, M., Desvoyes, B., Turina, M., Noad, R., and Scholthof, H. B. (2000). Genetic dissection of tomato bushy stunt virus p19-protein-mediated host-dependent symptom induction and systemic invasion. *Virology* **266:**79–87.

Clark, R. M., Schweikert, G., Toomajian, C., Ossowski, S., Zeller, G., Shinn, P., Warthmann, N., Hu, T. T., Fu, G., Hinds, D. A., Chen, H., Frazer, K. A., Huson, D. H., Scholkopf, B., Nordborg, M., Ratsch, G., Ecker, J. R., and Weigel, D. (2007). Common sequence polymorphisms shaping genetic diversity in *Arabidopsis thaliana*. *Science* **317:**338–342.

Collier, S. M., and Moffett, P. (2009). NB-LRRs work a "bait and switch" on pathogens. *Trends Plant Sci.* **14:**521–529.

Collmer, C. W., Marston, M. F., Taylor, J. C., and Jahn, M. (2000). The I gene of bean: A dosage-dependent allele conferring extreme resistance, hypersensitive resistance, or spreading vascular necrosis in response to the potyvirus Bean common mosaic virus. *Mol. Plant-Microbe Interact.* **13:**1266–1270.

Cooley, M. B., Pathirana, S., Wu, H. J., Kachroo, P., and Klessig, D. F. (2000). Members of the *Arabidopsis* HRT/RPP8 family of resistance genes confer resistance to both viral and oomycete pathogens. *Plant Cell* **12:**663–676.

Culver, J. N. (2002). Tobacco mosaic virus assembly and disassembly: Determinants in pathogenicity and resistance. *Annu. Rev. Phytopathol.* **40:**287–308.

Dangl, J. L., and Jones, J. D. (2001). Plant pathogens and integrated defence responses to infection. *Nature* **411:**826–833.

Dardick, C. D., Taraporewala, Z., Lu, B., and Culver, J. N. (1999). Comparison of tobamovirus coat protein structural features that affect elicitor activity in pepper, eggplant, and tobacco. *Mol. Plant-Microbe Interact.* **12:**247–251.

Diaz-Pendon, J. A., and Ding, S.-W. (2008). Direct and indirect roles of viral suppressors of RNA silencing in pathogenesis. *Annu. Rev. Phytopathol.* **46:**303–326.

Ding, S. W., and Voinnet, O. (2007). Antiviral immunity directed by small RNAs. *Cell* **130:**413–426.

Dodds, P. N., Lawrence, G. J., Catanzariti, A. M., Teh, T., Wang, C. I., Ayliffe, M. A., Kobe, B., and Ellis, J. G. (2006). Direct protein interaction underlies gene-for-gene specificity and coevolution of the flax resistance genes and flax rust avirulence genes. *Proc. Natl. Acad. Sci. U. S. A.* **103:**8888–8893.

Ellis, J. G., Dodds, P. N., and Lawrence, G. J. (2007a). Flax rust resistance gene specificity is based on direct resistance-avirulence protein interactions. *Annu. Rev. Phytopathol.* **45:**289–306.

Ellis, J. G., Lawrence, G. J., and Dodds, P. N. (2007b). Further analysis of gene-for-gene disease resistance specificity in flax. *Mol. Plant Pathol.* **8:**103–109.

Erickson, F. L., Holzberg, S., Calderon-Urrea, A., Handley, V., Axtell, M., Corr, C., and Baker, B. (1999). The helicase domain of the TMV replicase proteins induces the N-mediated defence response in tobacco. *Plant J.* **18:**67–75.

Ernst, K., Kumar, A., Kriseleit, D., Kloos, D. U., Phillips, M. S., and Ganal, M. W. (2002). The broad-spectrum potato cyst nematode resistance gene (Hero) from tomato is the only member of a large gene family of NBS-LRR genes with an unusual amino acid repeat in the LRR region. *Plant J.* **31:**127–136.

Farnham, G., and Baulcombe, D. C. (2006). Artificial evolution extends the spectrum of viruses that are targeted by a disease-resistance gene from potato. *Proc. Natl. Acad. Sci. U. S. A.* **103:**18828–18833.

Flor, H. H. (1971). Current status of gene-for-gene concept. *Annu. Rev. Phytopathol.* **9:**275–296.

Frost, D., Way, H., Howles, P., Luck, J., Manners, J., Hardham, A., Finnegan, J., and Ellis, J. (2004). Tobacco transgenic for the flax rust resistance gene L expresses allele-specific activation of defense responses. *Mol. Plant-Microbe Interact.* **17:**224–232.

Fujisaki, K., Hagihara, F., Azukawa, Y., Kaido, M., Okuno, T., and Mise, K. (2004). Identification and characterization of the SSB1 locus involved in symptom development by Spring beauty latent virus infection in *Arabidopsis thaliana*. *Mol. Plant-Microbe Interact.* **17:**967–975.

Garcia-Arenal, F., and McDonald, B. A. (2003). An analysis of the durability of resistance to plant viruses. *Phytopathology* **93:**941–952.

Gassmann, W., Dahlbeck, D., Chesnokova, O., Minsavage, G. V., Jones, J. B., and Staskawicz, B. J. (2000). Molecular evolution of virulence in natural field strains of *Xanthomonas campestris* pv. *vesicatoria*. *J. Bacteriol.* **182:**7053–7059.

Genda, Y., Kanda, A., Hamada, H., Sato, K., Ohnishi, J., and Tsuda, S. (2007). Two amino acid substitutions in the coat protein of pepper mild mottle virus are responsible for overcoming the L(4) gene-mediated resistance in *Capsicum* spp. *Phytopathology* **97:**787–793.

Gilardi, P., Garcia-Luque, I., and Serra, M. T. (2004). The coat protein of tobamovirus acts as elicitor of both L2 and L4 gene-mediated resistance in *Capsicum*. *J. Gen. Virol.* **85:**2077–2085.

Gimenez-Ibanez, S., Hann, D. R., Ntoukakis, V., Petutschnig, E., Lipka, V., and Rathjen, J. P. (2009). AvrPtoB targets the LysM receptor kinase CERK1 to promote bacterial virulence on plants. *Curr. Biol.* **19:**423–429.

Goggin, F. L., Jia, L., Shah, G., Hebert, S., Williamson, V. M., and Ullman, D. E. (2006). Heterologous expression of the Mi-1.2 gene from tomato confers resistance against nematodes but not aphids in eggplant. *Mol. Plant-Microbe Interact.* **19:**383–388.

Göhre, V., Spallek, T., Häweker, H., Mersmann, S., Mentzel, T., Boller, T., de Torres, M., Mansfield, J. W., and Robatzek, S. (2008). Plant pattern-recognition receptor FLS2 Is directed for degradation by the bacterial ubiquitin ligase AvrPtoB. *Curr. Biol.* **18:**1824–1832.

Grube, R. C., Radwanski, E. R., and Jahn, M. (2000). Comparative genetics of disease resistance within the Solanaceae. *Genetics* **155:**873–887.

Guo, M., Tian, F., Wamboldt, Y., and Alfano, J. R. (2009). The majority of the type III effector inventory of *Pseudomonas syringae* pv. tomato DC3000 can suppress plant immunity. *Mol. Plant-Microbe Interact.* **22:**1069–1080.

Hajimorad, M. R., Eggenberger, A. L., and Hill, J. H. (2005). Loss and gain of elicitor function of soybean mosaic virus G7 provoking Rsv1-mediated lethal systemic hypersensitive response maps to P3. *J. Virol.* **79:**1215–1222.

Harrison, B. D. (2002). Virus variation in relation to resistance-breaking in plants. *Euphytica* **124:**181–192.

Hayes, A. J., Jeong, S. C., Gore, M. A., Yu, Y. G., Buss, G. R., Tolin, S. A., and Maroof, M. A. (2004). Recombination within a nucleotide-binding-site/leucine-rich-repeat gene cluster produces new variants conditioning resistance to soybean mosaic virus in soybeans. *Genetics* **166:**493–503.

Howles, P., Lawrence, G., Finnegan, J., McFadden, H., Ayliffe, M., Dodds, P., and Ellis, J. (2005). Autoactive alleles of the flax L6 rust resistance gene induce non-race-specific rust resistance associated with the hypersensitive response. *Mol. Plant-Microbe Interact.* **18:**570–582.

Hsieh, Y. C., Omarov, R. T., and Scholthof, H. B. (2009). Diverse and newly recognized effects associated with short interfering RNA binding site modifications on the Tomato bushy stunt virus p19 silencing suppressor. *J. Virol.* **83:**2188–2200.

Hwang, C. F., Bhakta, A. V., Truesdell, G. M., Pudlo, W. M., and Williamson, V. M. (2000). Evidence for a role of the N terminus and leucine-rich repeat region of the Mi gene product in regulation of localized cell death. *Plant Cell* **12:** 1319–1329.

Iyer-Pascuzzi, A. S., and McCouch, S. R. (2007). Recessive resistance genes and the *Oryza sativa-Xanthomonas oryzae* pv. *oryzae* pathosystem. *Mol. Plant-Microbe Interact.* **20:**731–739.

Jeong, R. D., Chandra-Shekara, A. C., Kachroo, A., Klessig, D. F., and Kachroo, P. (2008). HRT-mediated hypersensitive response and resistance to Turnip crinkle virus in *Arabidopsis* does not require the function of TIP, the presumed guardee protein. *Mol. Plant-Microbe Interact.* **21:**1316–1324.

Jia, Y., McAdams, S. A., Bryan, G. T., Hershey, H. P., and Valent, B. (2000). Direct interaction of resistance gene and avirulence gene products confers rice blast resistance. *EMBO J.* **19:**4004–4014.

Jones, D. A., and Jones, J. D. G. (1997). The role of leucine-rich repeat proteins in plant defences. *Adv. Bot. Res.* **24:**89–167.

Kachroo, P., Yoshioka, K., Shah, J., Dooner, H. K., and Klessig, D. F. (2000). Resistance to turnip crinkle virus in *Arabidopsis* is regulated by two host genes and is salicylic acid dependent but NPR1, ethylene, and jasmonate independent. *Plant Cell* **12:**677–690.

Kaneko, Y. H., Inukai, T., Suehiro, N., Natsuaki, T., and Masuta, C. (2004). Fine genetic mapping of the TuNI locus causing systemic veinal necrosis by turnip mosaic virus infection in *Arabidopsis thaliana. Theor. Appl. Genet.* **110:**33–40.

Kim, Y. J., Lin, N. C., and Martin, G. B. (2002). Two distinct *Pseudomonas* effector proteins interact with the Pto kinase and activate plant immunity. *Cell* **109:**589–598.

Kohler, A., Rinaldi, C., Duplessis, S., Baucher, M., Geelen, D., Duchaussoy, F., Meyers, B. C., Boerjan, W., and Martin, F. (2008). Genome-wide identification of NBS resistance genes in *Populus trichocarpa. Plant Mol. Biol.* **66:**619–636.

Lanfermeijer, F. C., Dijkhuis, J., Sturre, M. J., de Haan, P., and Hille, J. (2003). Cloning and characterization of the durable tomato mosaic virus resistance gene Tm-2(2) from *Lycopersicon esculentum. Plant Mol. Biol.* **52:**1037–1049.

Lanfermeijer, F. C., Warmink, J., and Hille, J. (2005). The products of the broken Tm-2 and the durable Tm-2(2) resistance genes from tomato differ in four amino acids. *J. Exp. Bot.* **56:**2925–2933.

Lee, J. M., Hartman, G. L., Domier, L. L., and Bent, A. F. (1996). Identification and map location of TTR1, a single locus in *Arabidopsis thaliana* that confers tolerance to tobacco ringspot nepovirus. *Mol. Plant-Microbe Interact.* **9:**729–735.

Lee, S. W., Han, S. W., Bartley, L. E., and Ronald, P. C. (2006). From the Academy: Colloquium review. Unique characteristics of *Xanthomonas oryzae* pv. *oryzae* AvrXa21 and implications for plant innate immunity. *Proc. Natl. Acad. Sci. U. S. A.* **103:**18395–18400.

Leister, R. T., Dahlbeck, D., Day, B., Li, Y., Chesnokova, O., and Staskawicz, B. J. (2005). Molecular genetic evidence for the role of SGT1 in the intramolecular complementation of Bs2 protein activity in *Nicotiana benthamiana. Plant Cell* **17:**1268–1278.

Liu, J. J., and Ekramoddoullah, A. K. (2003). Isolation, genetic variation and expression of TIR-NBS-LRR resistance gene analogs from western white pine (*Pinus monticola* Dougl. ex. D. Don.). *Mol. Genet. Genomics* **270:**432–441.

Mackey, D., Holt, B. F., Wiig, A., and Dangl, J. L. (2002). RIN4 interacts with *Pseudomonas syringae* type III effector molecules and is required for RPM1-mediated resistance in *Arabidopsis*. *Cell* **108:**743–754.

Mackey, D., Belkhadir, Y., Alonso, J. M., Ecker, J. R., and Dangl, J. L. (2003). Arabidopsis RIN4 is a target of the type III virulence effector AvrRpt2 and modulates RPS2-mediated resistance. *Cell* **112:**379–389.

Martin, G. B., Bogdanove, A. J., and Sessa, G. (2003). Understanding the functions of plant disease resistance proteins. *Annu. Rev. Plant Biol.* **54:**23–61.

Mazourek, M., Cirulli, E. T., Collier, S. M., Landry, L. G., Kang, B.-C., Quirin, E. A., Bradeen, J. M., Moffett, P., and Jahn, M. (2009). The fractionated orthology of Bs2 and Rx/Gpa2 supports shared synteny of disease resistance in the Solanaceae. *Genetics* **182:**1351–1364.

McDonald, B. A., and Linde, C. (2002). The population genetics of plant pathogens and breeding strategies for durable resistance. *Euphytica* **124:**163–180.

McDowell, J. M., Dhandaydham, M., Long, T. A., Aarts, M. G., Goff, S., Holub, E. B., and Dangl, J. L. (1998). Intragenic recombination and diversifying selection contribute to the evolution of downy mildew resistance at the RPP8 locus of *Arabidopsis*. *Plant Cell* **10:**1861–1874.

McHale, L., Tan, X., Koehl, P., and Michelmore, R. W. (2006). Plant NBS-LRR proteins: Adaptable guards. *Genome Biol.* **7:**212.

Mestre, P., Brigneti, G., and Baulcombe, D. C. (2000). An Ry-mediated resistance response in potato requires the intact active site of the NIa proteinase from potato virus Y. *Plant J.* **23:**653–661.

Mestre, P., Brigneti, G., Durrant, M. C., and Baulcombe, D. C. (2003). Potato virus Y NIa protease activity is not sufficient for elicitation of Ry-mediated disease resistance in potato. *Plant J.* **36:**755–761.

Meyers, B. C., Shen, K. A., Rohani, P., Gaut, B. S., and Michelmore, R. W. (1998). Receptor-like genes in the major resistance locus of lettuce are subject to divergent selection. *Plant Cell* **10:**1833–1846.

Meyers, B. C., Dickerman, A. W., Michelmore, R. W., Sivaramakrishnan, S., Sobral, B. W., and Young, N. D. (1999). Plant disease resistance genes encode members of an ancient and diverse protein family within the nucleotide-binding superfamily. *Plant J.* **20:**317–332.

Meyers, B. C., Kozik, A., Griego, A., Kuang, H., and Michelmore, R. W. (2003). Genome-wide analysis of NBS-LRR-encoding genes in *Arabidopsis*. *Plant Cell* **15:**809–834.

Moffett, P and Klessig, D. F. (2008). Plant resistance to viruses: Natural resistance associated with dominant genes. *In* "Encyclopaedia of Virology" (B. W. J. Mahy and M. H. V. Van Regenmortel, eds.), 3rd edn, vol. 4, pp. 170–177. Elsevier, Oxford.

Moffett, P., Farnham, G., Peart, J., and Baulcombe, D. C. (2002). Interaction between domains of a plant NBS-LRR protein in disease resistance-related cell death. *EMBO J.* **21:**4511–4519.

Mondragon-Palomino, M., Meyers, B. C., Michelmore, R. W., and Gaut, B. S. (2002). Patterns of positive selection in the complete NBS-LRR gene family of *Arabidopsis thaliana*. *Genome Res.* **12:**1305–1315.

Mucyn, T. S., Clemente, A., Andriotis, V. M., Balmuth, A. L., Oldroyd, G. E., Staskawicz, B. J., and Rathjen, J. P. (2006). The tomato NBARC-LRR protein Prf interacts with Pto kinase in vivo to regulate specific plant immunity. *Plant Cell* **18:**2792–2806.

Mucyn, T. S., Wu, A. J., Balmuth, A. L., Arasteh, J. M., and Rathjen, J. P. (2009). Regulation of tomato Prf by Pto-like protein kinases. *Mol. Plant-Microbe Interact.* **22:**391–401.

Nemykh, M. A., Efimov, A. V., Novikov, V. K., Orlov, V. N., Arutyunyan, A. M., Drachev, V. A., Lukashina, E. V., Baratova, L. A., and Dobrov, E. N. (2008). One more probable structural transition in potato virus X virions and a revised model of the virus coat protein structure. *Virology* **373:**61–71.

Nicaise, V., Roux, M., and Zipfel, C. (2009). Recent advances in PAMP-triggered immunity against bacteria: Pattern recognition receptors watch over and raise the alarm. *Plant Physiol.* **150:**1638–1647.

Noel, L., Moores, T. L., van Der Biezen, E. A., Parniske, M., Daniels, M. J., Parker, J. E., and Jones, J. D. (1999). Pronounced intraspecific haplotype divergence at the RPP5 complex disease resistance locus of *Arabidopsis*. *Plant Cell* **11:**2099–2112.

Padgett, H. S., and Beachy, R. N. (1993). Analysis of a tobacco mosaic virus strain capable of overcoming N gene-mediated resistance. *Plant Cell* **5:**577–586.

Parlevliet, J. E. (2002). Durability of resistance against fungal, bacterial and viral pathogens; present situation. *Euphytica* **124:**147–156.

Pedley, K. F., and Martin, G. B. (2003). Molecular basis of Pto-mediated resistance to bacterial speck disease in tomato. *Annu. Rev. Phytopathol.* **41:**215–243.

Porter, B. W., Paidi, M., Ming, R., Alam, M., Nishijima, W. T., and Zhu, Y. J. (2009). Genome-wide analysis of *Carica papaya* reveals a small NBS resistance gene family. *Mol. Genet. Genomics* **281:**609–626.

Qu, F., Ren, T., and Morris, T. J. (2003). The coat protein of turnip crinkle virus suppresses posttranscriptional gene silencing at an early initiation step. *J. Virol.* **77:**511–522.

Querci, M., Baulcombe, D. C., Goldbach, R. W., and Salazar, L. F. (1995). Analysis of the resistance-breaking determinants of Potato-Virus-X (Pvx) strain Hb on different potato genotypes expressing extreme resistance to Pvx. *Phytopathology* **85:**1003–1010.

Rai, M. (2006). Refinement of the Citrus tristeza virus resistance gene (Ctv) positional map in *Poncirus trifoliata* and generation of transgenic grapefruit (*Citrus paradisi*) plant lines with candidate resistance genes in this region. *Plant Mol. Biol.* **61:**399–414.

Rairdan, G. J., and Moffett, P. (2006). Distinct domains in the ARC region of the potato resistance protein Rx mediate LRR binding and inhibition of activation. *Plant Cell* **18:**2082–2093.

Rairdan, G., and Moffett, P. (2007). Brothers in arms? Common and contrasting themes in pathogen perception by plant NB-LRR and animal NACHT-LRR proteins. *Microbes Infect.* **9:**677–686.

Rairdan, G. J., Collier, S. M., Sacco, M. A., Baldwin, T. T., Boettrich, T., and Moffett, P. (2008). The coiled-coil and nucleotide binding domains of the potato rx disease resistance protein function in pathogen recognition and signaling. *Plant Cell* **20:**739–751.

Ren, T., Qu, F., and Morris, T. J. (2000). HRT gene function requires interaction between a NAC protein and viral capsid protein to confer resistance to turnip crinkle virus. *Plant Cell* **12:**1917–1926.

Ren, T., Qu, F., and Morris, T. J. (2005). The nuclear localization of the *Arabidopsis* transcription factor TIP is blocked by its interaction with the coat protein of Turnip crinkle virus. *Virology* **331:**316–324.

Rentel, M. C., Leonelli, L., Dahlbeck, D., Zhao, B., and Staskawicz, B. J. (2008). Recognition of the *Hyaloperonospora parasitica* effector ATR13 triggeres resistance against oomycete, bacterial, and viral pathogens. *Proc. Natl. Acad. Sci. U. S. A.* **105:**1091–1096.

Robaglia, C., and Caranta, C. (2006). Translation initiation factors: A weak link in plant RNA virus infection. *Trends Plant Sci.* **11:**40–45.

Sacco, M. A. and Moffett, P. (2009). Disease resistance genes: Form and function. *In* "Molecular Plant-Microbe Interactions" (K. Bouarab, N. Brisson and F. Daayf, eds.), pp. 94–141. CABI, Wallingford, UK.

Sacco, M. A., Mansoor, S., and Moffett, P. (2007). A RanGAP protein physically interacts with the NB-LRR protein Rx, and is required for Rx-mediated viral resistance. *Plant J.* **52:**82–93.

Sacco, M. A., Koropacka, K., Grenier, E., Jaubert, M. J., Blanchard, A., Goverse, A., Smant, G., and Moffett, P. (2009). The cyst nematode SPRYSEC protein RBP-1 elicits Gpa2- and RanGAP2-dependent plant cell death. *PLoS Pathog.* **5:** e1000564.

Sakamoto, M., Tomita, R., Hamada, H., Iwadate, Y., Munemura, I., and Kobayashi, K. (2008). A primer-introduced restriction analysis-PCR-based method to analyse Pepper mild mottle virus populations in plants and field soil with respect to virus mutations that break L3 gene-mediated resistance of *Capsicum* plants. *Plant Pathology.* **57:**825–833.

Schoelz, J. E. (2006). Viral determinants of resistance and susceptibility. *In* "Natural Resistance Mechanisms of Plant to Viruses" (G. Loebenstein and J. Carr, eds.), pp. 13–43. Springer, Dordrecht, The Netherlands.

Scholthof, H. B., Scholthof, K. B., and Jackson, A. O. (1995). Identification of tomato bushy stunt virus host-specific symptom determinants by expression of individual genes from a potato virus X vector. *Plant Cell* **7:**1157–1172.

Shan, L., He, P., Li, J., Heese, A., Peck, S. C., Nürnberter, T., Martin, G. B., and Sheen, J. (2008). Bacterial effectors target the common signaling partner BAK1 to disrupt multiple MAMP receptor-signaling complexes and impede plant immunity. *Cell Host and Microbe* **4:**17–27.

Shao, F., Golstein, C., Ade, J., Stoutemyer, M., Dixon, J. E., and Innes, R. W. (2003). Cleavage of *Arabidopsis* PBS1 by a bacterial type III effector. *Science* **301:**1230–1233.

Shen, Q. H., Zhou, F., Bieri, S., Haizel, T., Shirasu, K., and Schulze-Lefert, P. (2003). Recognition specificity and RAR1/SGT1 dependence in barley Mla disease resistance genes to the powdery mildew fungus. *Plant Cell* **15:**732–744.

Staal, J., and Dixelius, C. (2007). Tracing the ancient origins of plant innate immunity. *Trends Plant Sci.* **12:**334–342.

Sun, Q., Collins, N. C., Ayliffe, M., Smith, S. M., Drake, J., Pryor, T., and Hulbert, S. H. (2001). Recombination between paralogues at the Rp1 rust resistance locus in maize. *Genetics* **158:**423–438.

Swiderski, M. R., Birker, D., and Jones, J. D. (2009). The TIR domain of TIR-NB-LRR resistance proteins is a signaling domain involved in cell death induction. *Mol. Plant-Microbe Interact.* **22:**157–165.

Takahashi, H., Suzuki, M., Natsuaki, K., Shigyo, T., Hino, K., Teraoka, T., Hosokawa, D., and Ehara, Y. (2001). Mapping the virus and host genes involved in the resistance response in cucumber mosaic virus-Infected *Arabidopsis thaliana. Plant Cell Physiol.* **42:**340–347.

Takahashi, H., Miller, J., Nozaki, Y., Takeda, M., Shah, J., Hase, S., Ikegami, M., Ehara, Y., and Dinesh-Kumar, S. P. (2002). RCY1, an *Arabidopsis thaliana* RPP8/HRT family resistance gene, conferring resistance to cucumber mosaic virus requires salicylic acid, ethylene and a novel signal transduction mechanism. *Plant J.* **32:**655–667.

Takken, F. L., Albrecht, M., and Tameling, W. I. (2006). Resistance proteins: Molecular switches of plant defence. *Curr. Opin. Plant Biol.* **9:**383–390.

Tameling, W. I., and Baulcombe, D. C. (2007). Physical association of the NB-LRR resistance protein Rx with a Ran GTPase-activating protein is required for extreme resistance to potato virus X. *Plant Cell* **19:**1682–1694.

Tameling, W. I., Elzinga, S. D., Darmin, P. S., Vossen, J. H., Takken, F. L., Haring, M. A., and Cornelissen, B. J. (2002). The tomato R gene products I-2 and MI-1 are functional ATP binding proteins with ATPase activity. *Plant Cell* **14:**2929–2939.

Tameling, W. I., Vossen, J. H., Albrecht, M., Lengauer, T., Berden, J. A., Haring, M. A., Cornelissen, B. J., and Takken, F. L. (2006). Mutations in the NB-ARC domain of I-2 that impair ATP hydrolysis cause autoactivation. *Plant Physiol.* **140:**1233–1245.

Thomas, C. L., Leh, V., Lederer, C., and Maule, A. J. (2003). Turnip crinkle virus coat protein mediates suppression of RNA silencing in *Nicotiana benthamiana. Virology* **306:**33–41.

Tobias, C. M., Oldroyd, G. E., Chang, J. H., and Staskawicz, B. J. (1999). Plants expressing the Pto disease resistance gene confer resistance to recombinant PVX containing the avirulence gene AvrPto. *The Plant J.* **17:**41–50.

Tomita, R., Murai, J., Miura, Y., Ishihara, H., Liu, S., Kubotera, Y., Honda, A., Hatta, R., Kuroda, T., Hamada, H., Sakamoto, M., Munemura, I., Nunomura, O., Ishikawa, K., Genda, Y., Kawasaki, S., Suzuki, K., Meksem, K., and Kobayashi, K. (2008). Fine mapping and DNA fiber FISH analysis locates the tobamovirus resistance gene L3 of *Capsicum chinense* in a 400-kb region of R-like genes cluster embedded in highly repetitive sequences. *Theor. Appl. Genet.* **117:**1107–1118.

Ueda, H., Yamaguchi, Y., and Sano, H. (2006). Direct interaction between the tobacco mosaic virus helicase domain and the ATP-bound resistance protein, N factor during the hypersensitive response in tobacco plants. *Plant Mol. Biol.* **61:**31–45.

Vallejos, C. E., Astua-Monge, G., Jones, V., Plyler, T. R., Sakiyama, N. S., and Mackenzie, S. A. (2006). Genetic and molecular characterization of the I locus of *Phaseolus vulgaris*. *Genetics* **172:**1229–1242.

van der Biezen, E. A., and Jones, J. D. (1998a). The NB-ARC domain: A novel signalling motif shared by plant resistance gene products and regulators of cell death in animals. *Curr. Biol.* **8:**R226–R227.

van der Biezen, E. A., and Jones, J. D. (1998b). Plant disease-resistance proteins and the gene-for-gene concept. *Trends Biochem. Sci.* **23:**454–456.

van der Hoorn, R. A. L., and Kamoun, S. (2008). From guard to decoy: A new model for perception of plant pathogen effectors. *Plant Cell* **20:**2009–2017.

van der Vossen, E. A., van der Voort, J. N., Kanyuka, K., Bendahmane, A., Sandbrink, H., Baulcombe, D. C., Bakker, J., Stiekema, W. J., and Klein-Lankhorst, R. M. (2000). Homologues of a single resistance-gene cluster in potato confer resistance to distinct pathogens: A virus and a nematode. *Plant J.* **23:**567–576.

van Loon, L. C. (1987). Disease induction by plant viruses. *Adv. Virus Res.* **33:**205–255.

Velasco, R., Zharkikh, A., Troggio, M., Cartwright, D. A., Cestaro, A., Pruss, D., Pindo, M., Fitzgerald, L. M., Vezzulli, S., Reid, J., Malacarne, G., Iliev, D., Coppola, G., Wardell, B., Micheletti, D., Macalma, T., Facci, M., Mitchell, J. T., Perazzolli, M., Eldredge, G., Gatto, P., Oyzerski, R., Moretto, M., Gutin, N., Stefanini, M., Chen, Y., Segala, C., Davenport, C., Dematte, L., Mraz, A., Battilana, J., Stormo, K., Costa, F., Tao, Q., Si-Ammour, A., Harkins, T., Lackey, A., Perbost, C., Taillon, B., Stella, A., Solovyev, V., Fawcett, J. A., Sterck, L., Vandepoele, K., Grando, S. M., Toppo, S., Moser, C., Lanchbury, J., Bogden, R., Skolnick, M., Sgaramella, V., Bhatnagar, S. K., Fontana, P., Gutin, A., Van de Peer, Y., Salamini, F., and Viola, R. (2007). A high quality draft consensus sequence of the genome of a heterozygous grapevine variety. *PLoS ONE* **2:**e1326.

Vidal, S., Cabrera, H., Andersson, R. A., Fredriksson, A., and Valkonen, J. P. T. (2002). Potato gene Y-1 is an N gene homolog that confers cell death upon infection with potato virus Y. *Mol. Plant-Microbe Interact.* **15:**717–727.

Weber, H., Schultze, S., and Pfitzner, A. J. (1993). Two amino acid substitutions in the tomato mosaic virus 30-kilodalton movement protein confer the ability to overcome the Tm-2(2) resistance gene in the tomato. *J. Virol.* **67:**6432–6438.

Weber, H., Ohnesorge, S., Silber, M. V., and Pfitzner, A. J. (2004). The Tomato mosaic virus 30 kDa movement protein interacts differentially with the resistance genes Tm-2 and Tm-2(2). *Arch. Virol.* **149:**1499–1514.

Whitham, S., Dinesh-Kumar, S. P., Choi, D., Hehl, R., Corr, C., and Baker, B. (1994). The product of the tobacco mosaic virus resistance gene N: Similarity to toll and the interleukin-1 receptor. *Cell* **78:**1101–1115.

Xiang, T., Zhong, N., Zou, Y., Wu, Y., Zhang, J., Xing, W., Li, Y., Tang, X., Zhu, L., Chai, J., and Zhou, J. M. (2008). *Pseudomonas syringae* effector AvrPto blocks innate immunity by targeting receptor kinases. *Curr. Biol.* **18:**74–80.

Yang, Z. N., Ye, X. R., Molina, J., Roose, M. L., and Mirkov, T. E. (2003). Sequence analysis of a 282-kilobase region surrounding the citrus Tristeza virus resistance gene (Ctv) locus in *Poncirus trifoliata* L. Raf. *Plant Physiol.* **131:**482–492.

Yoshimura, S., Yamanouchi, U., Katayose, Y., Toki, S., Wang, Z. X., Kono, I., Kurata, N., Yano, M., Iwata, N., and Sasaki, T. (1998). Expression of Xa1, a bacterial blight-resistance gene in rice, is induced by bacterial inoculation. *Proc. Natl. Acad. Sci. U. S. A.* **95:**1663–1668.

Zhou, J. M., and Chai, J. (2008). Plant pathogenic bacterial type III effectors subdue host responses. *Curr. Opin. Microbiol.* **11:**179–185.

Zhou, T., Wang, Y., Chen, J. Q., Araki, H., Jing, Z., Jiang, K., Shen, J., and Tian, D. (2004). Genome-wide identification of NBS genes in japonica rice reveals significant expansion of divergent non-TIR NBS-LRR genes. *Mol. Genet. Genomics* **271:**402–415.

Zhou, B., Qu, S., Liu, G., Dolan, M., Sakai, H., Lu, G., Bellizzi, M., and Wang, G. L. (2006). The eight amino-acid differences within three leucine-rich repeats between Pi2 and Piz-t resistance proteins determine the resistance specificity to *Magnaporthe grisea*. *Mol. Plant-Microbe Interact.* **19:**1216–1228.

Zipfel, C., and Rathjen, J. P. (2008). Plant immunity: AvrPto targets the frontline. *Curr. Biol.* **18:**R218–R220.

CHAPTER 2

RNA Silencing: An Antiviral Mechanism

T. Csorba[*], **V. Pantaleo**[†] and **J. Burgyán**[*,†]

Contents

Abstract

RNA silencing is an evolutionarily conserved sequence-specific gene-inactivation system that also functions as an antiviral mechanism in higher plants and insects. To overcome antiviral RNA silencing, viruses express silencing-suppressor proteins which can counteract the host silencing-based antiviral process. After the discovery of virus-encoded silencing suppressors, it was shown that these viral proteins can target one or more key points in the silencing machinery. Here we review recent progress in our understanding of the mechanism and function of antiviral RNA silencing in plants, and on the virus's counterattack by expression of silencing-suppressor proteins. We also discuss emerging

* Agricultural Biotechnology Center, Plant Biology Institute, P.O. Box 411, H-2101 Gödöllõ, Hungary
† Istituto di Virologia Vegetale del CNR, Torino, Italy

Advances in Virus Research, Volume 75
ISSN: 0065-3527, DOI: 10.1016/S0065-3527(09)07502-2

evidence that RNA silencing and expression of viral silencing-suppressor proteins are tools forged as a consequence of virus–host coevolution for fine-tuning host–pathogen coexistence.

I. INTRODUCTION

Viruses are obligate intracellular pathogens that infect almost all living organisms. They are mostly composed of two, sometimes three components: the viral genome of either DNA or RNA, a protein coat, which protects the genome, and a not obligatory third part, the envelope, which surrounds the virus particles originating mostly from the host (e.g. Rhabdoviridae, Tospoviridae) (Matthews, 1991). Subviral RNAs such as satellite RNAs (satRNAs), defective RNAs (D-RNAs) or defective interfering RNAs (DI-RNAs) are frequently found associated with plant viruses; these RNAs can be distinguished from the viral genome by their dispensability for normal virus propagation. Subviral RNA replication is completely dependent on enzymes encoded by their helper virus and thus amplification is limited to coinfected cells (Simon *et al.*, 2004). Plants are also occasionally infected with viroids, the smallest self-replicating plant pathogens known to date. Their genomes consist of short, naked, circular, single-stranded RNA with a high degree of secondary structure, and without any protein coding capacity (Flores *et al.*, 2005).

The genomes of plant viruses show huge diversity having genomes of DNA, RNA, linear, circular or segmented, single- or double-stranded, positive (+), negative (−) or ambisense (+/−). These differences between viral genomes imply differences in the respective viral replication strategies. RNA viruses encode their own RNA-dependent RNA polymerase (RdRp). The DNA genome of pararetroviruses is replicated by reverse transcription involving RNA and DNA intermediates, through the action of reverse transcriptase encoded by the viral genome; viroids use host DNA-dependent RNA polymerases and replicate via RNA intermediates (Hull, 2002).

The presence and replication of viruses in the host induce diverse mechanisms for combating viral infection at the level of single cells and the whole organism. These mechanisms range from RNA interference a mechanism mainly found in plants and lower eukaryotes, to the sophisticated interferon-regulated gene response of higher animals. Since all types of viruses at a given point of their replication reach the stage of ssRNA or dsRNA, they actively provoke RNA-induced silencing-based host defense responses (Ding and Voinnet, 2007).

RNA silencing relies on small RNA (sRNA) molecules, approximately 21–24 nucleotides long, so-called short interfering RNAs (siRNAs) and micro RNAs (miRNAs) (Hamilton and Baulcombe, 1999; Hamilton *et al.*,

2002; Kim, 2005; Plasterk, 2002). Biochemical and genetic analyses have shown that the core mechanisms of RNA silencing are shared among different eukaryotes (Baulcombe, 2004; Hannon and Conklin, 2004; Meister and Tuschl, 2004; Plasterk, 2002; Voinnet, 2002; Zamore, 2002). RNA silencing is triggered by double-stranded (ds) or self-complementary foldback RNAs that are processed into 21–24 nt short siRNA or miRNA duplexes by the RNase III-type DICER enzymes (Bartel, 2004; Baulcombe, 2004; Bernstein *et al.*, 2001). These miRNAs and siRNAs activate a multiprotein effector complex, the RNA-induced silencing complex (RISC) (Hammond *et al.*, 2000; Tomari and Zamore, 2005), of which Argonaute protein (AGO) is the slicer component showing similarity to RNase H (Liu *et al.*, 2004a; Song *et al.*, 2004; Tomari and Zamore, 2005). RISC is the executioner of RNA silencing, inhibiting target RNA expression. The specific recognition of target sequences is guided by the sRNAs through a base-pairing mechanism, whereas the slicing of target RNA is carried out by the AGO proteins at the post-transcriptional or transcriptional levels (Almeida and Allshire, 2005; Bartel, 2004; Brodersen *et al.*, 2008; Eamens, *et al.*, 2008).

Short RNAs can also guide another effector complex, namely the RNA-induced transcriptional gene silencing (RITS) complex to direct the chromatin modification of homologous DNA sequences (Verdel *et al.*, 2004).

RNA silencing regulates several biological processes via down-regulation of gene expression by miRNAs and siRNAs such as developmental timing and patterning, transposon control, DNA methylation and chromatin modification as well as antiviral defense.

One of the best-established functions of RNA silencing is antiviral defense, which was first discovered in plants (Dougherty *et al.*, 1994; Lindbo *et al.*, 1993; Ratcliff *et al.*, 1997). The antiviral functions of RNA silencing are supported by the following observations: first, virus-derived siRNAs (viRNAs) accumulate at high level during viral infections and can effectively target the viral RNA. Second, most if not all plant viruses have evolved virulence factors called viral suppressors of RNA silencing (VSRs) to overcome the RNA silencing-based host defense.

II. RNA-BASED ANTIVIRAL IMMUNITY

The first indications that RNA-mediated responses play an important antiviral role came from observations that transgenic expression of viral sequences protected plants from homologous viruses by conferring a sequence-specific degradation of challenging viral RNAs (Dougherty *et al.*, 1994; Lindbo and Dougherty, 1992). Later it was shown that viruses are potentially both initiators and targets of gene silencing (Ratcliff *et al.*, 1997). Subsequently, it has been shown that several viruses encode proteins, which suppress RNA silencing-mediated defense (Voinnet *et al.*, 1999) indicating that these pathogens have evolved counter-defensive strategies against RNA silencing.

A. Mounting the antiviral defense

The main steps of mounting antiviral silencing are: (i) activation of RNA silencing in the cell by the incoming viral RNA, where structured or double-stranded RNA molecules are recognized by plant Dicer-like (DCL) enzymes, producing vsiRNAs; (ii) the protection of vsiRNAs by 2′ O-methylation. These vsiRNAs are then recruited by AGO-containing complexes to target cognate viral RNAs. Alternatively these vsiRNAs can enter the plant RNA-dependent RNA polymerase (RDR)-mediated amplification cycle to enhance the antiviral silencing response (Fig. 1).

1. Activation of RNA silencing and production of vsiRNAs

The majority of known plant viruses have RNA genomes and replicate via double-stranded replication intermediates, at first suggesting that these molecules are the main trigger of RNA silencing. However it turned out that induction of the silencing response is much more complex. The probability that viral RNAs are present in a naked form in the plant cell is very small. The majority of viral RNAs are in encapsidated form or in complexes for replication or movement. Moreover viral replication usually takes place inside a specialized replication compartment and the viral dsRNA replication intermediates can immediately be unwound by viral or host RNA helicases (Ahlquist, 2002). It is more likely that the highly structured single-stranded viral RNAs with stem-loop structures are recognized by the silencing machinery, and the double-stranded regions directly chopped by plant DCLs into virus-derived siRNAs (vsiRNAs) (Fig. 1). The sequencing and experimental data of vsiRNAs strongly support this model since the resulting vsiRNA molecules are imperfect duplexes (Molnar *et al.*, 2005) that have a non-random distribution along the viral genome, and they map asymmetrically to the positive strand of the viral RNA (Donaire *et al.*, 2009; Ho *et al.*, 2006; Molnar *et al.*, 2005; Qi *et al.*, 2009; Szittya *et al.*, unpublished results). Similarly, in the case of *Cauliflower mosaic virus* (CaMV) the 35S polycistronic transcript of this dsDNA virus contains an extensive secondary structure, which is the major vsiRNA source (Moissiard and Voinnet, 2006). In viroid-infected plants the strandness of viroid-specific siRNAs is also asymmetrical and they are preferentially derived from the highly structured plus sense viroid RNA sequence (Itaya *et al.*, 2007), although recent deep sequencing data have shown more symmetrical origin of viroid siRNAs (Navarro *et al.*, unpublished results). Furthermore, in plants infected by the *Potyvirus Turnip mosaic virus* (TuMV),

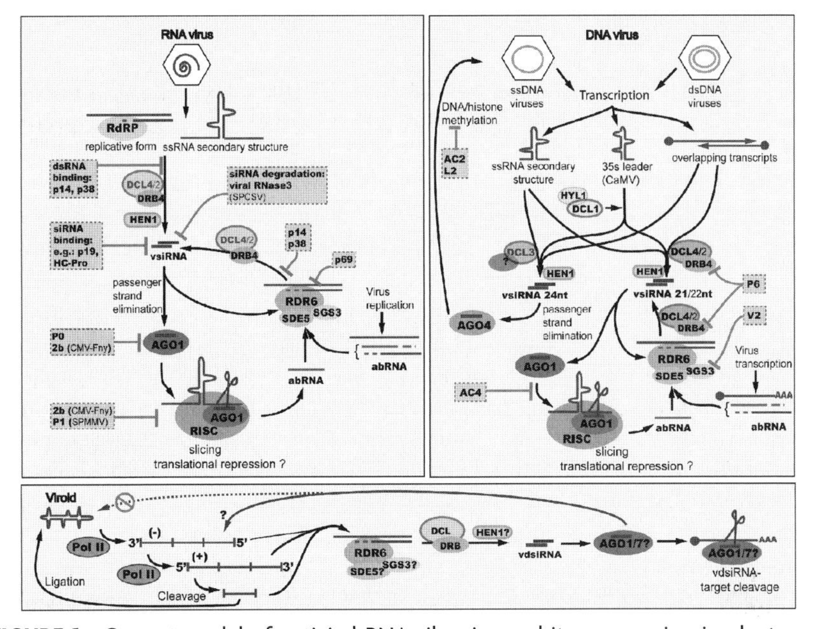

FIGURE 1 Current model of antiviral RNA silencing and its suppression in plants. RNA silencing is initiated by the perception of viral dsRNA or partially double-stranded hairpin RNA, which are processed to 21 nt viral siRNAs (vsiRNAs) by dsRNA-specific RNases called Dicer like 4 (DCL4) in association with dsRNA-binding protein 4 (DRB4). If DCL4 is suppressed or inactive DCL2 can replace it, generating 22 nt vsiRNAs. The vsiRNA are stabilized by 2′-O-methylation by HUA ENHANCER1 (HEN1) and afterward incorporated into Argonaute1 (AGO1) protein, the major antiviral slicer. Other members of the AGO family like AGO7 may be also involved. Plant RISC may also inhibit viral gene expression by translational arrest, although this has not yet been proved to be an active antiviral mechanism. In addition to incorporation into RISC, may vsiRNAs also take part in amplification of the silencing response through the action of RNA-dependent RNA polymerase 6 (RDR6) and its cofactors. The viral RNA molecules cleaved by RISC and viral RNAs lacking 5′- or/and 3′- end are likely recognized being aberrant RNAs (abRNA) and converted to dsRNA by RDR6 action. This dsRNA processed again vsiRNAs, that leads to generation of more vsiRNA and amplification of the silencing response. In DNA virus infections all four DCLs are involved in the production of 21–23 nt long vsiRNAs. In the case of circular ssDNA geminiviruses highly structured regions of viral ssRNA transcripts or dsRNA molecules formed by overlapping complementary viral RNA transcripts recognized by DCL3/4/2 and processed vsiRNAs. DCL1 has only a very limited role in ssDNA virus-derived vsiRNA production (Blevins *et al.*, 2006). In dsDNA virus infection such as CaMV RNA silencing triggered mainly by the highly structured 35S leader sequence of viral mRNA transcript. DCL3 and DCL4, are the most important dicers implicated in the production of vsiRNAs derived from CaMV transcripts. DCL2 activity is evident especially when DCL4 is inactive. DCL1 has a facilitating role, possibly making the 35S leader sequence more accessible for the other dicers (Moissiard and Voinnet, 2006). The DCL3-dependent 24 nt vsiRNAs are incorporated into AGO4, and may direct DNA/histone methylation of the DNA virus genome in the nucleus. DCL1- and DCL4-dependent 21 nt vsiRNAs are recruited by AGO1 to direct slicing or are implicated in RDR6 pathway-mediated amplification (Ding and Voinnet, 2007). (see Page 3 in Color Section at the back of the book.)

which has a positive ssRNA genome that expresses a polyprotein, the sequenced vsiRNAs showed similar amounts of (+) and (−) strand vsiRNAs (Ho *et al.*, 2006). This result may suggest that the TuMV derived vsiRNAs are processed from dsRNA.

In the case of circular ssDNA geminiviruses a part of vsiRNAs are likely derived from dsRNAs formed by overlapping sense–antisense transcripts (Akbergenov *et al.*, 2006; Blevins *et al.*, 2006; Ding and Voinnet, 2007). These findings demonstrate that the perfect dsRNAs can also be a substrate for vsiRNAs indicating that plant DCLs are adapted to different viral replication and expression strategies and are able to recognize the different RNA structures, which are formed during virus life cycles.

The *Arabidopsis thaliana* genome encodes four DCLs for sRNA processing: DCL1 to DCL4. Specific DCLs have major functions in specific silencing pathways but functional redundancy exists between members: DCL1 contributes to miRNA production and has no or little role in the antiviral response. DCL2, DCL3 and DCL4 are able to recognize viral structures and, respectively, generate vsiRNAs of 22, 24 and 21 nt in length (Blevins *et al.*, 2006; Deleris *et al.*, 2006).

Biogenesis of vsiRNAs needs the coordinated and hierarchical action of DCL enzymes (Moissiard and Voinnet, 2006). RNA virus infection is mainly affected by DCL4 and to a lesser extent by DCL2. Inactivation of DCL4 reveals the subordinate antiviral role of DCL2. Deactivation by mutation of both DCL2 and DCL4 was necessary and sufficient to restore systemic infection of a suppressor-deficient virus, indicating the crucial role of DCL4 and DCL2 in the antiviral response (Bouche *et al.*, 2006; Deleris *et al.*, 2006).

Upon DNA virus infection, the production of 24 nt vsiRNA by DCL3 is also sufficient for virus-induced gene silencing (Blevins *et al.*, 2006). DCL3-dependent 24-nt long vsiRNAs have also been detected in *Tobacco rattle virus* (TRV) and *Cucumber mosaic virus* (CMV) infected wild-type (wt) plants or *Turnip crinkle virus* (TCV) infected *dcl4/dcl2* double mutant *Arabidopsis* plants (Deleris *et al.*, 2006; Qu *et al.*, 2008).

The participation of DCL1 in the antiviral silencing induced by RNA viruses is slight since DCL1-dependent vsiRNAs are hardly detected in the *dcl2/dcl3/dcl4* triple mutant plants (Blevins *et al.*, 2006; Bouche *et al.*, 2006; Deleris *et al.*, 2006). However, DCL1 promotes DCL3- and DCL4-derived vsiRNA accumulation upon dsDNA (CaMV) or ssDNA (geminiviruses) infection. Very likely DCL1 excises the stem-loop structures of 35S leader transcripts, which are very similar to pre- or pri-miRNAs and renders them more accessible to other DCLs (Moissiard and Voinnet, 2006). An opposite effect of DCL1 was found in plants infected with TCV: the disruption of DCL1 function led to higher expression of DCL4 and DCL3, and enhanced antiviral response, suggesting that these proteins are under DCL1-negative control (Qu *et al.*, 2008).

Plant dsRNA-binding proteins (DRBs) have been found associated with DCLs, facilitating their production of sRNAs (Vaucheret, 2006) (Fig. 1). In *Arabidopsis* plants, five DRBs have been identified. While DCLs act redundantly and hierarchically, there is little or no redundancy or hierarchy among the DRBs in their DCL interactions. HYPONASTIC LEAVES1 (HYL1) is a DRB protein that cooperates with DCL1 and is required in processing of miRNA precursors in the plant cell nucleus (Hiraguri *et al.*, 2005). DCL4 operates exclusively with DRB4 to produce trans-acting (ta) siRNAs (Adenot *et al.*, 2006; Nakazawa *et al.*, 2007) and 21 nt siRNAs from viral RNAs (Hiraguri *et al.*, 2005). DRB proteins associated with DCL2 and DCL3 are likely involved in vsiRNAs and natural siRNAs generation.

Whether DRBs are also associated with heterochromatic siRNA production has not yet been reported. Co-localization of DCL1 and HYL1 or DCL4 and DRB4 partners suggests that they could form heterodimer complexes (Hiraguri *et al.*, 2005). In *drb4* mutant plants, a high level of silencing-suppressor mutant TCV-ΔCP RNA was detected compared to wt plants, but less than in *dcl4* plants, and the vsiRNA accumulation of TCV-ΔCP in *drb4* mutant plants was slightly decreased compared to wt plants. These findings suggest that DRB4 may not be involved directly in vsiRNA production but rather in vsiRNA stabilization or delivery to effector complexes (Qu *et al.*, 2008).

2. Protection of sRNAs by 3′ end methylation

The biogenesis of sRNA in plants requires an additional step apart from DCL-mediated processing (Fig. 1). This is a methylation reaction catalyzed by HUA ENHANCER 1 (HEN1) methyltransferase, which links a methyl group to the ribose of the 3′ last nucleotide of the sRNA duplex in a sequence-independent manner (Yu *et al.*, 2005). The 2′-O-methylation of 3′end appears to protect sRNA molecules against uridylation (Li *et al.*, 2005) and against the exoribonuclease activity of small RNA degrading nucleases (SDN1-3) (Ramachandran and Chen, 2008). All types of endogenous sRNAs are methylated in plants whereas in insects and vertebrates only the germline-specific piRNAs are methylated (Ohara *et al.*, 2007). Resistance to β-elimination has proved that plant virus-derived vsiRNAs are also methylated (Akbergenov *et al.*, 2006; Blevins *et al.*, 2006; Csorba *et al.*, 2007; Lozsa *et al.*, 2008).

Hen1 mutant plants accumulate less vsiRNAs from both RNA and DNA viruses and exhibit reduced levels of silencing (Blevins *et al.*, 2006). HEN1 is the only methyltransferase involved in methylation of vsiRNAs and miRNAs (Csorba *et al.*, 2007). Experiments using siRNA-binding suppressors suggest that vsiRNAs are methylated in the cytoplasm while miRNAs are methylated in the nucleus and in the cytoplasm (Lozsa *et al.*, 2008). The finding that the p122 suppressor expressed by cr-TMV could interfere only partially with miRNA methylation supports this scenario. This also suggests that miRNAs are exported from the nucleus in both methylated and non-methylated forms. Strikingly the methylation of miRNAs could not be inhibited in cr-TMV infected HASTY mutant plants (*hst-15*), where the export of miRNA from nucleus to cytoplasm is compromised (Csorba *et al.*, 2007). In line with this observation HEN1 is reported to be present in both the nucleus and the cytoplasm (Fang and Spector, 2007).

B. Effector steps of antiviral silencing

1. Antiviral RNA-induced silencing complexes

The *Arabidopsis* genome contains ten AGO proteins, AGO1 to AGO10, and they are the catalytic components of RNA silencing effector complexes. They interact with small RNAs to effect gene silencing in all RNAi-related pathways known so far (Fig. 1). AGO proteins are characterized by two principal domains: the sRNA-binding PAZ domain at the N-terminus (Ma *et al.*, 2004) and the PIWI domain with its metal-coordinating DDE catalytic triad at the C-terminus, responsible for RNaseH-like "slicer" activity on target ssRNAs complementary to the sRNA loaded within the AGO (Tolia and Joshua-Tor, 2007). The functional equivalent in HsAGO2 contains the DDH motif (Rivas *et al.*, 2005). The presence of the catalytic triad does not necessarily imply slicer activity, indeed miRNA-loaded AGOs can silence gene expression through translational arrest without slicing (Bartel, 2004).

The DCL-mediated processing of viral dsRNA regions into vsiRNA in theory could be enough for viral RNA degradation. However, *dcl2/dcl3*, *dcl2/dcl4* and *dcl3/dcl4* mutant plants infected with TRV had approximately equivalent levels of vsiRNAs but only *dcl2/dcl4* plants showed strong viral symptoms and high virus titer (Deleris *et al.*, 2006) suggesting that dicing *per se* is not sufficient for defense against virus infection, and additional effector complex action is required.

AGO1 was suggested to be involved in antiviral silencing, as hypomorphic *ago1* mutants are hypersensitive to CMV infection (Morel *et al.*, 2002). Pull-down experiments revealed that AGO1 recruits miRNAs, tasiRNAs, transgene-derived siRNAs and that AGO1-sRNA complex had slicer activity *in vitro* (Baumberger and Baulcombe, 2005; Qi *et al.*, 2005). Subsequently Zhang *et al.* (2006) have shown that AGO1 also recruits vsiRNAs and the AGO1–vsiRNA complex is a major player in antiviral defense. In addition, very recent studies demonstrated that both AGO2 and AGO5 can bind CMV-derived vsiRNAs, selecting for short RNAs having 5′- A and C nucleotides, respectively (Mi *et al.*, 2008; Takeda *et al.*, 2008).

More direct evidence of the existence of antiviral RISC comes from studies with the positive-strand RNA *Cymbidium ringspot virus*. Two vsiRNA-containing silencing complexes, which co-fractionated with miRNA-containing complexes were detected in infected plants: the smaller one at approximately the AGO1-siRNA size (150 kDa), the so-called minimal-RISC, and a high molecular weight (670 kDa) multiprotein complex (Pantaleo *et al.*, 2007) probably homologous to animal RISC (Pham *et al.*, 2004). A similar complex was isolated in separate experiments involving another tombusvirus. This complex contained vsiRNAs and exhibited *in vitro* nuclease activity, which preferentially targeted homologous viral sequences (Omarov *et al.*, 2007). Strikingly, viral RNA was targeted in a non-random fashion in hotspots by the antiviral RISC in Cym19stop suppressor mutant virus-infected plants (Pantaleo *et al.*, 2007).

Those regions of viral RNA that show hotspots for vsiRNA generation probably form strong secondary structures, which are selectively recognized by DCLs. However, these hotspots are poor targets for RISC-mediated cleavage, since RNA sequences possessing strong secondary structures are not accessible for RISC (Ameres *et al.*, 2007; Pantaleo *et al.*, 2007; Szittya *et al.*, 2002). The accessibility of the viral RNA is probably also influenced by encapsidation, formation of replication complexes containing host and viral proteins and compartmentalization of virus replication. It has been shown recently that there is asymmetry in the strandness of virus-derived siRNAs, showing that the majority of viral siRNAs have plus-stranded viral sequences (Donaire *et al.*, 2009; Ho *et al.*, 2006; Molnar *et al.*, 2005; Qi *et al.*, 2009; Szittya *et al.*, unpublished results). This finding suggests that viral siRNA-guided RISC should target more frequently the viral strand having negative polarity than the plus-stranded viral RNA. Indeed, in previous experiments strand-specific sensors were used for sensing antiviral RISC-mediated cleavages and the sensor RNAs carrying (−) strand sequences were better target than the (+) strand sensors (Pantaleo *et al.*, 2007). It is worth noting that the amount of negative-strand viral RNA is a rate-limiting factor for viral replication; thus preferential targeting of the negative viral strand makes the antiviral silencing response very efficient and very attractive for plant defense. The analysis of 5′-RNA cleavage products of sensor RNAs and viral RNAs reveals the presence of non-templated U residues at the cleavage site (Pantaleo *et al.*, 2007), this is the signature of RISC action (Shen and Goodman, 2004), confirming the presence of RISC-mediated slicing.

According to the current model of virus-induced RNA silencing (Fig. 1) a large amount of vsiRNA originates from partially base-paired regions of plus-stranded viral RNAs (Ding and Voinnet, 2007; Molnar *et al.*, 2005; Szittya *et al.*, unpublished results). Thus plus-stranded vsiRNAs could also potentially target plus-stranded viral RNA through translational arresting. Indeed, recent findings suggest that translational arresting could also be a widespread way to inhibit gene expression by plant miRNAs and siRNAs (Brodersen *et al.*, 2008; Lanet *et al.*, 2009). Moreover, a novel role of AGO4 has been suggested for specific translational control of viral RNA (Bhattacharjee *et al.*, 2009). AGO7 was shown to function as a surrogate slicer in the absence of AGO1 in the clearance of viral RNA of TCV, and favors less structured RNA targets (Qu *et al.*, 2008).

Another possibility for antiviral defense occurs at the transcriptional level, and is encountered with DNA viruses. *De novo* asymmetric cytosine methylation occurs on *Tomato leaf curl virus* DNA and restricts its replication (Alberter *et al.*, 2005; Bian *et al.*, 2006).

2. Amplification of silencing response

The third family of proteins involved in silencing in plants is the RDR family. In plants there are six RDR paralogs: RDR1, RDR2, RDR3a (RDR3), RDR3b (RDR4), RDR3c (RDR5) and RDR6 (SDE1/SGS-2). The putative catalytic domain is the DLDGD motif, which is highly conserved among all RDRs identified (Wassenegger and Krczal, 2006). In the silencing pathways RDRs synthesize cRNA from the 3′- terminal nucleotides of the template RNA. Then the template and the cRNA remain bound forming a perfectly base-paired dsRNA molecule, which is later processed by DCLs into siRNAs.

Plant RDRs have important homeostatic and defensive functions. The major cellular function of RDR is its involvement in the trans-acting siRNA biogenesis. The process is initiated by miRNA-directed cleavage of non-coding trans-acting siRNA primary transcripts (TAS) (Allen *et al.*, 2005; Howell *et al.*, 2007) and the cleaved TAS RNA is converted to dsRNA by RDR6. The resulting dsRNA is processed by DCL4 to in-phase 21 nt tasiRNAs, which regulate endogenous targets that may control organ development and juvenile-to-adult transition (Hunter *et al.*, 2006).

The other important role of RDRs is defense against selfish and foreign nucleic acids like transposons, transgenes or viruses through the amplification and spreading of RNA silencing. In plants, amplification of the silencing response occurs in at least two different ways (Fig. 1). In the priming-dependent mechanism, viral or transgene-derived primary siRNAs recruit RDRs to the cognate ssRNA, which is converted to dsRNA through synthesis of complementary RNA. This dsRNA is then processed to secondary siRNA by DCLs (Voinnet, 2005). Plant RDRs can also amplify the silencing response in a primer-independent manner, in which RDRs detect the somehow aberrant (different from normal cellular and viral RNAs) RNA molecules deriving from viruses, transgenes or transposons, convert it into dsRNA which becomes the substrate for DCLs, and produce secondary siRNAs. Recent studies have demonstrated experimentally the generation and accumulation of secondary vsiRNAs in plants infected with CMV (Diaz-Pendon *et al.*, 2007). These siRNAs, upon incorporation into RISC complexes, execute effector steps of silencing and also direct further amplification rounds by releasing the cleaved target RNAs, additional templates for RDR enzymes (Vaucheret, 2006; Voinnet, 2005). These vsiRNA were able to act both in cell-autonomous and non-cell-autonomous fashion (Dunoyer *et al.*, 2005).

De novo dsRNA synthesis mediated by the host RDR pathway may play an important role in antiviral silencing against some viruses such as CMV, since *Arabidopsis* mutants lacking components of the AGO1-RDR6-SGS3-SDE5 pathway show enhanced disease susceptibility (Vaucheret, 2006; Voinnet, 2005). Tobacco plants in which RDR6 activity was silenced are also hypersusceptible to several unrelated (+) ssRNA viruses (Qu *et al.*, 2005; Schwach *et al.*, 2005). However, this is not a general phenomenon for all plant viruses since other studies showed that loss-of-function mutations in RDR6 have no detectable impact on the production of vsiRNAs and virus accumulation in *Arabidopsis* plants infected with TRV, TCV and cr-TMV (Blevins *et al.*, 2006; Dalmay *et al.*, 2000, 2001; Deleris *et al.*, 2006).

The existence of six RDRs suggests redundancy and specialization between RDRs in the different pathways. The nuclear-localized RDR2 is involved in DCL3-AGO4 dependent heterochromatic silencing (Matzke and Birchler, 2005), and RDR1 has a role in defense against tobamo-viruses, tobraviruses and potexviruses (Yang *et al.*, 2004; Yu *et al.*, 2003). RDR1 is strongly induced by salicylic acid (Xie *et al.*, 2001), a defense-signaling hormone, whereas RDR6 expression is controlled by the stress hormone abscisic acid (Yang *et al.*, 2008). Recently RDR2 was also implicated in the antiviral defense against TRV, where the major contributors are RDR1 and RDR6 (Donaire *et al.*, 2008). vsiRNA production is strongly reduced in triple *rdr1/rdr2/rdr6* mutants, pointing to the importance of RDR action in generating substrates for DCLs.

It has also been suggested that the RDR-dependent secondary vsiRNAs can drive a more effective antiviral response, against some but not all virus infections (Vaistij and Jones, 2009). These findings indicate that although there are very conserved steps in the silencing-based antiviral response, plants are able to respond specifically to different viruses, demonstrating the versatility of this antiviral surveillance mechanism. Plants attacked by viruses can thus activate alternative pathways to counteract the invasion with an appropriate strategy; this system has likely evolved to face the many different viruses possessing their ample portfolio of replication, infection, transmission and silencing suppression strategies.

III. SILENCING SUPPRESSION STRATEGIES

A decade ago the discovery of VSRs provided the most convincing evidence for the antiviral nature of RNA silencing and revealed the pathogen counter-defensive strategy of active suppression of host surveillance (Voinnet *et al.*, 1999). More than 50 individual VSRs have been identified from almost all plant virus genera (Table 1), underlining the need of their expression for successful virus infection (Diaz-Pendon and Ding, 2008; Ding and Voinnet, 2007). Available data suggest that virtually all plant viruses encode at least one suppressor, but in many cases viruses encode more than one (e.g. carmo-, clostero-, crini- and begomoviruses; see Table 1).

Viral suppressors are considered to be of recent evolutionary origin, often encoded by out-of-frame ORFs within more ancient genes. They are surprisingly diverse within and across kingdoms with no obvious sequence homology (Ding and Voinnet, 2007). VSRs are variously positioned on the viral genome and expressed using different strategies such as subgenomic RNAs, transcriptional read-through, ribosomal leaky-scanning or proteolytic maturation of polyproteins. Due to their evolution many of the suppressors identified to date are multifunctional: beside being RNA-silencing suppressors they also perform essential roles by functioning as coat protein, replicase, movement protein, helper-component for virus transmission, protease or transcriptional regulators. Virtually all steps of the silencing pathway have been found to be targeted by VSRs; either acting on silencing-related RNA molecules or through protein–protein interaction (Fig. 1; Table 1).

A. Suppressors targeting silencing-related RNAs

The most widely used suppression strategy, adopted by many viral genera (tospo-, cucumo-, poty-, ipomo-, tombus-, clostero-, viti-, tobamo- and hordeiviruses) is ds siRNA sequestration (Lakatos *et al.*, 2006; Merai *et al.*, 2006), which prevents assembly of the RISC effector complex (see Table 1 and the references within). Importantly, these siRNA-binding VSRs are completely unrelated proteins although they share analogous biochemical properties, suggesting their independent evolution in different viruses.

siRNA binding is exemplified by the tombusvirus p19 protein, probably the most studied viral silencing suppressor so far. Crystallographic studies have shown that the head-to-tail p19 homodimer acts like a molecular caliper, which measures the length of siRNAs and binds them with high affinity in a sequence-independent way selecting for the 19 bp long dsRNA duplex region of the typical siRNA (Vargason *et al.*, 2003; Ye *et al.*, 2003). P19 demonstrates extraordinary adaptation of a viral protein to inactivate vsiRNAs, which are the most conserved key element of the antiviral silencing response. Other VSRs such as the *Cucumovirus Tomato aspermy virus* (TAV) 2b protein or B2 of the insect-infecting *Flock House virus* also show siRNA-binding activity, however structural studies have shown that the structures of silencing-suppressor proteins and their mode of binding

TABLE 1 Identified silencing suppressor proteins encoded by plant viruses

Family	Genus	Type species	Supressor	Suppression mechanisms	Other functions	References
dsRNA						
Reoviridae	*Phytoreovirus*	*Rice dwarf virus*	**Pns10**	Upstream to dsRNA	Unknown	Cao *et al.* (2005)
ss (−) RNA						
Bunyaviridae	*Tospovirus*	*Tomato spotted wilt virus*	**NSs**	Inhibition of sense-PTGS	Pathogenicity determinant	Bucher *et al.* (2003), Takeda *et al.* (2002)
No family	*Tenuivirus*	*Rice hoja blanca virus*	**NS3**	siRNA binding	Unknown	Hemmes *et al.* (2007)
		Rice stripe virus	**NS3**	ss-,ds-siRNA and ssRNA-binding	Unknown	Xiong *et al.* (2009)
ss (+) RNA						
Bromoviridae	*Cucumovirus*	*Cucumber mosaic virus* (Fny)	**2b**	AGO1 interaction	Host specific movement	Zhang *et al.* (2006)
		Cucumber mosaic virus (CM95R)	**2b**	siRNA binding	Host specific movement	Goto *et al.* (2007)
		Tomato aspermy virus	**2b**	siRNA binding	Host specific movement	Chen *et al.* (2008)
Comoviridae	*Comovirus*	*Cowpea mosaic virus*	**S protein**	Unknown	Small coat protein	Canizares *et al.* (2004), Liu *et al.* (2004b)
Potyviridae	*Potyvirus*	*Potato virus Y*	**HC-Pro**	siRNA binding	Movement, polyprotein processing	Kasschau and Carrington (1998)

Table 1 *(Continued)*

Family	Genus	Type species	Supressor	Suppression mechanisms	Other functions	References
		Tobacco etch virus	**HC-Pro**	siRNA binding	Aphid transmission, pathogenicity determinant	Lakatos *et al.* (2006)
	Ipomovirus	*Cassava brown streak virus*	**P1**	Unknown	Serine proteinase	Mbanzibwa *et al.* (2009)
		Sweet potato mild mottle virus	**P1**	AGO1 interaction	Serine proteinase	Giner *et al.* (unpublished results)
		Cucumber vein yellowing virus	**P1b**	siRNA binding	Serine proteinase	Valli *et al.* (2008)
Tombusviridae	*Tombusvirus*	*Carnation Italian ringspot virus*	**p19**	siRNA binding	Movement, pathogenicity determinant	Silhavy *et al.* (2002), Vargason *et al.* (2003)
		Tobacco bushy stunt virus	**p19**	siRNA binding	Movement, pathogenicity determinant	Voinnet *et al.* (1999)
	Aureusvirus	*Pothos latent virus*	**p14**	dsRNA binding	Pathogenicity determinant	Merai *et al.* (2005)
	Carmovirus	*Turnip crinkle virus*	**p38**	dsRNA binding	Coat protein	Merai *et al.* (2006), Thomas *et al.* (2003)
		Melon necrotic spot virus	**p7B**	Unknown	Movement	Genoves *et al.* (2006)
			p42	Unknown	Coat protein, pathogenicity determinant	Genoves *et al.* (2006)

		Hibiscus chlorotic ringspot virus	**CP**	Downstream to RDR6	Coat protein	Meng *et al.* (2008)
	Dianthovirus	*Red clover necrotic mosaic virus*	**replication**	Host factor sequestration, e.g. DCL1	Replication	Takeda *et al.* (2005)
			MP	Unknown	Movement	Powers *et al.* (2008)
		**Satellite panicum mosaic virus*			**CP** suppressor of VSR	Qiu and Scholthof (2004)
Closteroviridae	*Closterovirus*	*Beet yellows virus*	**P21**	siRNA binding	Replication enhancer	Lakatos *et al.* (2006), Reed *et al.* (2003)
		Citrus tristeza virus	**p20**	Unknown	Replication enhancer	Lu *et al.* (2004)
			p23	Unknown	Nucleic acid binding	Lu *et al.* (2004)
			CP	Unknown	Coat protein	Lu *et al.* (2004)
	Crinivirus	*Sweet potato chlorotic stunt virus*	**RNase3**	siRNA cleavage	Pathogenicity determinant, synergism	Cuellar *et al.* (2009), Kreuze *et al.* (2005)
			p22	Unknown	Pathogenicity determinant	Cuellar *et al.* (2008)
		Cucurbit yellow stunting disorder virus	**p25**	Unknown	Unknown	Kataya *et al.* (2009)
		Tomato chlorosis virus	**p22**	Suppress local RNA silencing	Unknown	Canizares *et al.* (2008)
			CP	Unknown	Coat protein	Canizares *et al.* (2008)

Table 1 *(Continued)*

Family	Genus	Type species	Supressor	Suppression mechanisms	Other functions	References
			CPm	Unknown	Coat protein minor	Canizares *et al.* (2008)
Luteoviridae	*Polerovirus*	*Beet western yellows virus*	**P0**	AGO destabilization	Pathogenicity determinant	Baumberger *et al.* (2007),Bortolamiol *et al.* (2007)
Tymoviridae	*Tymovirus*	*Turnip yellow mosaic virus*	**p69**	Upstream to dsRNA formation	Movement, pathogenicity determinant	Chen *et al.* (2004)
Flexiviridae	*Potexvirus*	*Potato virus X*	**P25**	Inhibits systemic silencing	Movement	Voinnet *et al.* (2000)
	Trichovirus	*Apple chlorotic leafspot virus*	**p50**	Inhibits long distant movement of silencing	Movement	Yaegashi *et al.* (2007,2008)
	Vitivirus	*Grapevine virus A*	**p10**	ss-,ds-siRNA binding	RNA-binding, movement, pathogenicity determinant	Chiba *et al.* (2006), Zhou *et al.* (2006)
No family	*Tobamovirus*	*Tobacco mosaic virus*	**p126**	siRNA binding	Replicase subunit	Harries *et al.* (2008)
		Cr-Tobacco mosaic virus	**p122**	siRNA binding	Replicase subunit	Csorba *et al.* (2007)
		Tomato mosaic virus	**p130**	siRNA binding	Replicase subunit	Kubota *et al.* (2003)
	Tobravirus	*Tobacco rattle virus*	**16K**	Downstream to dsRNA	Unknown	Martinez-Priego *et al.* (2008)

	Furovirus	*Soil-borne wheat mosaic virus*	**19K**	Systemic silencing inhibition	Pathogenicity determinant	Te *et al.* (2005)
	Pecluvirus	*Peanut clump virus*	**P15**	siRNA binding	Intercellular virus movement	Dunoyer *et al.* (2002)
	Benyvirus	*Beet necrotic yellow vein virus*	**p31 (roots)**	Inhibits silencing in roots	Vector transmission, pathogenicity determinant	Rahim *et al.* (2007)
			p14	Unknown	Regulation of RNA2 and CP expression	Dunoyer *et al.* (2002), Rahim *et al.* (2007)
	Hordeivirus	*Barley stripe mosaic virus*	**γB**	siRNA binding	Pathogenicity determinant	Merai *et al.* (2006), Yelina *et al.* (2002)
	Sobemovirus	*Rice yellow mottle virus*	**P1**	Unknown	Movement, pathogenicity determinant, virus accumulation	Sire *et al.* (2008), Voinnet *et al.* (1999)
ssDNA						
Geminiviridae	*Curtovirus*	*Beet curly top virus*	**L2**	Inhibits ADK and transmethylation	Pathogenicity determinant	Wang *et al.* (2003, 2005)
	Begomovirus	*African cassava mosaic virus*	**AC4**	ssRNA binding	Movement, pathogenicity determinant	Bisaro (2006), Chellappan *et al.* (2005)
			AC2		Transcriptional transactivator	Voinnet *et al.* (1999)
		Tomato golden mosaic virus	**AL2**	Inhibits ADK and transmethylation	Synergistic genes: AC2–AC4	Wang *et al.* (2005)

Table 1 *(Continued)*

Family	Genus	Type species	Supressor	Suppression mechanisms	Other functions	References
		Mungbean yellow mosaic virus	**AC2**	Activates endogenous silencing suppressor		Trinks *et al.* (2005)
		Tomato yellow leaf curl virus	**V2**	Inhibits SGS3 activity	Unknown	Fukunaga and Doudna (2009), Glick *et al.* (2008)
		**Satellite DNAβ*	**βC1**	Unknown	Replication, movement	Saunders *et al.* (2004))
dsDNA (RT)						
Caulimoviridae	*Caulimovirus*	*Cauliflower mosaic virus*	**P6**	RDB4 interaction	Translational transactivator	Haas *et al.* (2008), Love *et al.* (2007)

siRNAs do not share any similarity (Chao *et al.*, 2005; Chen *et al.*, 2008). Recently, it was also reported that siRNA-binding suppressors (p19, HC-Pro, p122) may prevent the essential siRNA and miRNA 2′-O-methylation steps in the biogenesis of siRNA and miRNA (Csorba *et al.*, 2007; Ebhardt *et al.*, 2005; Lozsa *et al.*, 2008; Vogler *et al.*, 2007). However, it seems that this inhibitory effect requires temporal and spatial co-expression of the suppressor, endogenous or viral siRNAs and miRNAs (Lozsa *et al.*, 2008). In the presence of siRNA-binding VSRs, plants fail to confine the infection and virus spread occurs, since vsiRNAs are sequestered by the VSRs before they can be incorporated in silencing effector complexes.

A very similar outcome is achieved by adopting a completely different strategy in the case of the *Crinivirus Sweet potato chlorotic stunt virus* (SPCSV). SPCSV-encoded RNase3 endonuclease cleaves the 21-, 22- and 24-vsiRNAs into 14 bp products, which are inactive in the RNA-silencing pathways (Cuellar *et al.*, 2009). The p14 of *Pothos latent aureusvirus* and p38 of *Turnip crinkle virus* (TCV) are potent VSRs and bind long and short dsRNAs (including ds siRNAs) in a size-independent way (Merai *et al.*, 2005, 2006). p14 and p38 may interact with the ds viral RNA, inhibiting the RNA-silencing machinery on two levels: (i) by siRNA sequestration (Merai *et al.*, 2006), and (ii) by interfering with DCL4-mediated vsiRNA processing. The inhibition of DCL4 by p38 has also been confirmed experimentally (Deleris *et al.*, 2006). In contrast to dsRNA-binding VSRs, the AC4 suppressor of *African cassava mosaic virus* binds single-stranded small RNAs bound by AGOs and prevents holo RISC assembly. AC4 also inhibits miRNA-mediated negative regulation of endogenous genes (Chellappan *et al.*, 2005). *Rice stripe virus* NS3 and *Grapevine virus A* p10 proteins are also able to sequester ss-siRNA molecules (Xiong *et al.*, 2009; Zhou *et al.*, 2006) implying, in part at least, a similar strategy to AC4.

Previous studies have shown that the V2 protein from *Tomato yellow leaf curl virus* is an efficient suppressor of RNA silencing (Glick *et al.*, 2008; Zrachya *et al.*, 2007) and V2 was proposed to interact with the tomato protein SGS3 (SlSGS3) in infected plant cells (Glick *et al.*, 2008). However, recent *in vitro* studies on V2 show that it outcompetes SGS3 protein for binding a dsRNA with 5′ ssRNA overhangs, whereas a V2 mutant lacking the suppressor function *in vivo* cannot efficiently overcome SGS3 binding (Fukunaga and Doudna, 2009). These findings not only predict a new type of RNA-binding silencing suppressor but also may reveal a new RNA intermediate, which is essential for SDS3/RDR6-dependent siRNA formation in the plant (Kumakura *et al.*, 2009).

B. Suppressors interacting with silencing-related host proteins

The 2b protein of CMV was one of the first VSRs described (Brigneti *et al.*, 1998). In plants efficient virus infection requires the inhibition of either the short or long-range silencing signal of antiviral RNA silencing. 2b prevents the spread of the long-range silencing signal, and so facilitate the systemic virus infection (Guo and Ding, 2002). Indeed, the 2b-deficient mutant virus (CMV-Δ2b) replicates in tobacco protoplasts at wt level but its accumulation is 20-fold lower in inoculated tobacco leaves and it is not detectable in the upper leaves (Soards *et al.*, 2002). In inoculated leaves CMV-Δ2b infects plant cells in small isolated spots, whereas wt CMV infects over large areas. CMV-Δ2b can be rescued by the *dcl2/dcl4* host double mutant, which is impaired in vsiRNA production, indicating that 2b is dispensable for infection and spread in a host defective in small RNA-directed immunity (Diaz-Pendon *et al.*, 2007). RDR-dependent CMV vsiRNA production is strongly reduced in the presence of 2b (Diaz-Pendon *et al.*, 2007). This suggests that 2b facilitates short- and long-distance virus spread but in the absence of 2b plant tissues can set up their antiviral machinery, which restricts further spreading of the virus.

Consistently, 2b of Fny-CMV has been found to physically interact on PAZ and part of the PIWI domain with siRNA-loaded AGO1, and inhibits its slicing activity (Zhang *et al.*, 2006). Fny-CMV 2b was found to colocalize with AGO1 protein preferentially in the cell's nucleus but also in cytoplasmic foci (Mayers *et al.*, 2000). Fny-CMV 2b protein expression phenocopies the *ago1-27* mutant phenotype and leads to the accumulation the inactive miRNA duplexes (formed by mature miRNA and the normally labile passenger strand, called miRNA*), and miRNA-target accumulation (Zhang *et al.*, 2006). The phenotype of Fny-CMV 2b expressing transgenic plants is similar to plants expressing other siRNA-binding suppressors (Dunoyer *et al.*, 2004; Lewsey *et al.*, 2007; Zhang *et al.*, 2006).

Chen *et al.* (2008) reported that 2b of TAV a cucumovirus related to CMV binds siRNA duplexes. Analysis of the crystal structure of TAV-2b-siRNA has shown that 2b adopts an alpha-helix structure to form a homodimer, and binds to siRNA by measuring its length, similarly to tombusvirus p19, although, the structures of the two VSRs (p19 and 2b) do not share any similarity. 2b of the severe CMV strain CM95R is also known to bind siRNAs (Goto *et al.*, 2007). Thus, cucumovirus 2b could have a dual mode of action, either sequestering siRNAs or interacting with AGO1.

As recently described, the 29 kDa P0 protein of the phloem-limited poleroviruses targets Argonautes, the core component of RISC for degradation (Baumberger *et al.*, 2007; Bortolamiol *et al.*, 2007; Pazhouhandeh *et al.*, 2006). This protein is indispensable for viral infection. Null mutations of P0 in *Beet western yellows virus* (BWYV) and *Potato leafroll virus* strongly diminish or completely abolish virus accumulation (Mayo and Ziegler-Graff, 1996). In contrast to the RNA-binding VSRs, P0 has no

RNA-binding activity (Zhang *et al.*, 2006) (Csorba *et al.*, unpublished results); instead it interacts with the SCF family of E3-ligase SKP1 (S-phase kinase-related protein 1) components orthologous to *Arabidopsis* ASK1 and ASK2, by means of its minimal F-box motif and promotes Argonaute degradation. Disruption of the F-box motif by mutation annuls P0 silencing-suppressor activity. Downregulation of SKP homologues in *Nicotiana benthamiana* plants by virus-induced gene silencing leads to resistance against BWYV infection (Pazhouhandeh *et al.*, 2006).

The P0 is suggested to interact with PAZ and adjacent upstream domains of multiple Argonautes (AGO1, AGO2, AGO4-6, AGO9), however this interaction is probably transient or indirect *in vivo*. AGO degradation seems to be 26S proteasome-independent (Baumberger *et al.*, 2007) probably involving other cellular proteases. Transgenic expression of P0 in *Arabidopsis* leads to severe developmental abnormalities similar to those induced by mutants affecting miRNA pathways, which is accompanied by AGO1 protein decay *in planta* and enhanced levels of several miRNA-target transcripts (Bortolamiol *et al.*, 2007).

Earlier results suggested that the impact of P0 on plant endogenous silencing pathways is so devastating that it is unfavorable even for the virus itself (Pfeffer *et al.*, 2002). In natural virus infection P0 expression is limited by a suboptimal translation initiation codon. Attempts to optimize the translation initiation region have failed: backward mutations restored the poor translation initiation codon characteristic to the wt or ended up in additional mutation creating a termination codon downstream (Pfeffer *et al.*, 2002), showing that P0 overexpression is unfavorable for the virus.

Interestingly, large amounts of polyubiquitinated proteins accumulate upon BWYV P0 ectopic expression (Csorba *et al.*, unpublished results), suggesting that BWYV P0 may have multiple targets in the cell or induces protein-based immunity; this points to a link between RNA silencing and protein-based defense strategies. This idea is supported by the fact that transient expression of P0 induces a dose-dependent cell death phenotype similar to that caused by the P0 of *Sugarcane yellow leaf virus*, another polerovirus (Mangwende *et al.*, 2009).

A new type of AGO-interacting VSR was recently characterized in our laboratory. The P1 suppressor of the *Ipomovirus Sweet potato mild mottle virus*, seems to act by inhibition of siRNA and miRNA programmed RISC through targeting AGO1. Suppression activity was mapped to the N-terminal part of P1, a region containing WG/GW motifs essential both for AGO binding and for suppression (Giner *et al.*, unpublished results). The conserved GW182 family of proteins has recently been identified and the family members have been shown to be associated with miRISC and to be required for miRNA-mediated gene silencing. Proteins containing WG/GW motifs have been found in animals and plants where they are thought to bind AGOs and be required for proper RISC function. In animals, the GW182 proteins, such as P-body components, have been found essential for miRNA-induced silencing and mRNA degradation (Behm-Ansmant *et al.*, 2006; Eulalio *et al.*, 2008; Liu *et al.*, 2005). The plant RNA polymerase IVb also contains several WG/GW motifs, which are required for AGO4 binding and RNA-directed DNA methylation (RdDM) (El-Shami *et al.*, 2007). Recently, KTF1 protein containing WG/GW motifs and SPT5-like domains has been identified as AGO4-binding proteins playing an important role in RdDM (Bies-Etheve *et al.*, 2009; He *et al.*, 2009). Thus, the action of P1 represents a novel mode of RNA-silencing suppression, which might act by out-competing cellular components with similar motifs, and that this is radically different from other VSR mechanisms described.

C. Other silencing suppressor strategies

The p69 protein of the positive-strand RNA *Turnip yellow mosaic virus* (TYMV) suppresses RNA silencing induced by sense–transgenes (S-PTGS) but not silencing induced by inverted-repeat transgenes (IR-PTGS) (Chen *et al.*, 2004); the negative-strand RNA virus *Tomato spotted wilt virus* (TSWV) encodes a silencing suppressor, the NSs protein, which appears to adopt a mechanistically similar strategy. In a transient co-expression assay, NSs suppresses local and systemic S-PTGS, but not IR-PTGS (Takeda *et al.*, 2002). This suggests that these suppressors could interfere with dsRNA generation by inhibition of plant RDRs or other components of this pathway. Consistent with these observations is the fact that p69 expression leads to a phenotype characteristic for *rdr6* mutant (Chen *et al.*, 2004; Dalmay *et al.*, 2000).

Suppressors from the *Geminiviridae* family nicely exemplify that silencing suppressors may modulate endogenous biochemical pathways for virus benefit. The *Tomato golden mosaic virus* (TGMV)-encoded AL2 protein and the closely related *Beet curly top virus* (BCTV) L2 interact with and inactivate adenosine kinase (ADK), a cellular enzyme important for adenosine salvage and the methyl cycle. Plants infected with the *l2* mutant BCTV and other unrelated viruses display increased ADK activity, suggesting that ADK could be part of a plant response to virus infection (Wang *et al.*, 2003). ADK has a role in sustaining the methylation cycle. By inhibiting ADK, the AL2 and L2 proteins indirectly block this cycle, and thus could interfere with epigenic modification of the viral genome (Bisaro, 2006; Wang *et al.*, 2005). *In vitro* methylated TGMV cannot replicate in protoplasts (Bisaro, 2006), suggesting that the methylation of the viral genome could be a valid mode to combat geminivirus infection.

Evidence concerning the transcription-dependent activity of *Mungbean yellow mosaic virus* (MYMV) and *African cassava mosaic virus* (ACMV) protein AC2 has also been obtained. AC2 induces expression of more than 30 host genes, including *Werner exonuclease-like 1* (*WEL1*) an endogenous negative regulator of silencing (Trinks *et al.*, 2005). The picture is more complex since these genes also include positive regulators of silencing. AC4 of ACMV but not that of *East African cassava mosaic virus* was suggested to bind ss-siRNAs and miRNAs. Thus AC4 uses a novel mechanism different from that of AC2, to block silencing by interfering with RISC loading downstream to ds sRNA unwinding.

AC4 expression in transgenic plants leads to severe developmental defects since miRNA pathway is also disrupted (Chellappan *et al.*, 2005). In the presence of AC4 the level of miRNA targets is upregulated, but miRNA level is downregulated. This implies that AC4-mediated sequestration of ss-sRNA has a different outcome to that of the siRNA- and miRNA-binding suppressors, where the RNA duplexes are stabilized by the suppressors. The different geminiviral AC2 and AC4 proteins are not equally efficient in suppressing silencing, and the presence of two different mechanisms may explain in part the synergy observed in mixed geminivirus infections (Vanitharani *et al.*, 2004).

Host factors involved in both RNA-silencing suppression and viral replication have been proposed to play roles in RNA-silencing suppression during infection by *Red clover necrotic mosaic virus* (RCNMV). Upon RCMV infection there is a close relationship between negative-strand RNA synthesis and RNA-silencing suppression. It has been suggested that sequestration of host factors required for antiviral silencing could reduce the silencing response. The putative host factor involved in both processes could be DCL1 protein, since miRNA biogenesis is inhibited by virus replication and *dcl1* mutant plants show reduced susceptibility to RCNMV infection (Takeda *et al.*, 2005). In the suggested scenario, DCL1 and its homologues are recruited by the viral replication complex and therefore depleted from the silencing pathways.

The above examples show that plant viruses have evolved various strategies to counteract antiviral RNA-silencing mechanisms. The majority of silencing suppression strategies target conserved key elements of RNA-silencing pathways such as siRNAs or their precursors and crucial enzymes like AGO proteins; sometimes a single VSR can target more than one element in the silencing pathways.

The large variety of well-described VSRs also offers better understanding of plant silencing pathways through targeting specific steps of silencing machinery.

IV. SILENCING SUPPRESSORS AND VIRAL SYMPTOMS

Viral infection leads to various symptoms such as development of lesions, dark green islands and growth defects (Hull, 2002). Although many VSRs (Table 1) have been identified as pathogenic determinants largely responsible for virus-induced symptoms, the molecular basis for virus-induced diseases in plants has been a long-standing mystery. It is well established that the antiviral and endogenous silencing pathways share common elements, and VSRs have been shown to interfere with those pathways. siRNA-binding VSRs (e.g. HC-Pro and p19) could interact with siRNA and miRNA biogenesis at different stages. This interference may alter endogenous gene expression regulated through miRNAs or siRNAs. Similarly, long dsRNA-binding VSRs (e.g. p38 and p14) could compromise DCLs or AGO1-targeting VSRs (e.g. P0 and P1) inhibit RISC activities, which in turn may alter the expression of an unpredictable number of genes involved in plant development. Indeed expression of VSRs in transgenic plants leads to phenotypes that mimic virus symptoms (Chapman *et al.*, 2004; Dunoyer *et al.*, 2004; Kasschau *et al.*, 2003).

However, transgenic expression of VSRs does not necessary reflect the effects of viral infection on endogenous silencing pathways, since in natural viral infection, expression of VSRs is restricted to virus-infected tissues and compartments, and is also limited in time. In fact recent results show that inhibition of 3′ modification of vsiRNAs and miRNAs in virus-infected plants requires spatial and temporal co-expression of small RNAs and VSRs (Lozsa *et al.*, 2008).

V. CONCLUDING REMARKS

During the last few years dramatic progress has been made in understanding the roles and pathways involved in antiviral RNA silencing. A large number of new silencing-suppressor proteins have been described from almost all plant virus genera. The discovery of the molecular bases of silencing suppression for many proteins has inspired new concepts on the existence of cellular negative regulators of RNA silencing, such as silencing suppressors. In virus-infected plants the key function of RNA silencing is to protect plants against viral invasion. Surprisingly it seems that viruses may exploit this defense to keep the virus titer at a tolerable level in plant tissues through controlling the expression level of VSRs. For example in natural virus infection a suboptimal codon controls the expression of the polerovirus P0 VSR, thus the moderate inhibition of RNA silencing ensures that both the viruses and the plants survive (Pfeffer *et al.*, 2002).

It is likely that antiviral RNA silencing accelerates the continuous modification/evolution of viral genome since even a single base change in the target site of antiviral si/miRISC could protect the viral genome against degradation. However, this protection is very temporary since the vsiRNAs produced from the modified new sequence can target the viral genome again. Therefore, this is a continuous selection pressure for the RNA genome to alter the si/miRISC target site sequence. To escape from this endless circle, viruses evolved VSRs to protect their genome. Alternatively, viruses evolved their genome to be highly structured, which is not accessible for RISC. For example the fast evolution of CymRSV DI-RNAs ended up with a short highly structured DI-RNA, which are resistant against RNA silencing (Szittya *et al.*, 2002). The highly structured rod-like form of matured viroid genome is another example for the structure-mediated resistance of a RNA molecule to RNA silencing (Gomez *et al.*, 2009). On the other hand highly structured RNA molecules are good substrates for plant DCLs. Indeed, silencing-resistant DI-RNAs of CymRSV efficiently trigger RNA silencing against their helper genomes, while the generated vsiRNAs are not able to target the highly structured DI-RNAs (Szittya *et al.*, 2002).

The fast evolution of viral genome under RNA silencing pressure was also exemplified by introducing natural or artificial miRNA target site in the viral genome (Lin *et al.*, 2009; Simon-Mateo and Garcia, 2006). The most common outcome was the deletion or modification of the target site in the viral genome. The fast mutation of the viral genome may explain why host plant derived miRNAs or siRNAs are not found to target viral genome in natural virus resistance.

An extraordinary adaptation of viruses to the antiviral silencing has been found in the CymRSV–satellite RNA system. It has been shown that the helper virus harnesses RNA-silencing mechanism to control the accumulation of the virus parasitic satellite RNA (Pantaleo and Burgyan, 2008). Thus, RNA silencing appears to be involved in many ways in this fine-tuning of plant–virus interplay for joint survival, but our knowledge is still limited about the regulation of this intimate plant–virus interaction, which remains for future exploration.

ACKNOWLEDGMENTS

We thank Robert Geoffrey Milne for critical reading of the manuscript and for his valuable comments. VP and JB are supported by bilateral research program between Consiglio Nazionale delle Ricerche (CNR, Italy) and Magyar Tudományos Akadémia (MTA, Hungary). TCS and JB are funded by the European Commission (FP6 Integrated Project SIROCCO LSHG-CT-2006-037900).

REFERENCES

Adenot, X., Elmayan, T., Lauressergues, D., Boutet, S., Bouche, N., Gasciolli, V., and Vaucheret, H. (2006). DRB4-dependent TAS3 trans-acting siRNAs control leaf morphology through AGO7. *Curr. Biol.* **16:**927–932.

Ahlquist, P. (2002). RNA-dependent RNA polymerases, viruses, and RNA silencing. *Science* **296:**1270–1273.
Akbergenov, R., Si-Ammour, A., Blevins, T., Amin, I., Kutter, C., Vanderschuren, H., Zhang, P., Gruissem, W., Meins, F. Jr., Hohn, T., and Pooggin, M. M. (2006). Molecular characterization of geminivirus-derived small RNAs in different plant species. *Nucleic Acids Res.* **34:**462–471.
Alberter, B., Ali Rezaian, M., and Jeske, H. (2005). Replicative intermediates of Tomato leaf curl virus and its satellite DNAs. *Virology* **331:**441–448.
Allen, E., Xie, Z., Gustafson, A. M., and Carrington, J. C. (2005). microRNA-directed phasing during trans-acting siRNA biogenesis in plants. *Cell* **121:**207–221.
Almeida, R., and Allshire, R. C. (2005). RNA silencing and genome regulation. *Trends Cell Biol.* **15:**251–258.
Ameres, S. L., Martinez, J., and Schroeder, R. (2007). Molecular basis for target RNA recognition and cleavage by human RISC. *Cell* **130:**101–112.
Bartel, D. P. (2004). MicroRNAs: genomics, biogenesis, mechanism, and function. *Cell* **116:**281–297.
Baulcombe, D. (2004). RNA silencing in plants. *Nature* **431:**356–363.
Baumberger, N., and Baulcombe, D. C. (2005). Arabidopsis ARGONAUTE1 is an RNA Slicer that selectively recruits microRNAs and short interfering RNAs. *Proc. Natl. Acad. Sci. U. S. A.* **102:**11928–11933.
Baumberger, N., Tsai, C. H., Lie, M., Havecker, E., and Baulcombe, D. C. (2007). The polerovirus silencing suppressor P0 targets argonaute proteins for degradation. *Curr. Biol.* **17:**1609–1614.
Behm-Ansmant, I., Rehwinkel, J., Doerks, T., Stark, A., Bork, P., and Izaurralde, E. (2006). mRNA degradation by miRNAs and GW182 requires both CCR4:NOT deadenylase and DCP1:DCP2 decapping complexes. *Genes Dev.* **20:**1885–1898.
Bernstein, E., Caudy, A. A., Hammond, S. M., and Hannon, G. J. (2001). Role for a bidentate ribonuclease in the initiation step of RNA interference. *Nature* **409:**363–366.
Bhattacharjee, S., Zamora, A., Azhar, M. T., Sacco, M. A., Lambert, L. H., and Moffett, P. (2009). Virus resistance induced by NB-LRR proteins involves Argonaute4-dependent translational control. *Plant J.* **58:**940–951.
Bian, X. Y., Rasheed, M. S., Seemanpillai, M. J., and Ali Rezaian, M. (2006). Analysis of silencing escape of tomato leaf curl virus: an evaluation of the role of DNA methylation. *Mol. Plant-Microbe Interact.* **19:**614–624.
Bies-Etheve, N., Pontier, D., Lahmy, S., Picart, C., Vega, D., Cooke, R., and Lagrange, T. (2009). RNA-directed DNA methylation requires an AGO4-interacting member of the SPT5 elongation factor family. *EMBO Rep.* **10:**649–654.
Bisaro, D. M. (2006). Silencing suppression by geminivirus proteins. *Virology* **344:**158–168.
Blevins, T., Rajeswaran, R., Shivaprasad, P. V., Beknazariants, D., Si-Ammour, A., Park, H. S., Vazquez, F., Robertson, D., Meins, F. Jr., Hohn, T., and Pooggin, M. M. (2006). Four plant Dicers mediate viral small RNA biogenesis and DNA virus induced silencing. *Nucleic Acids Res.* **34:**6233–6246.

Bortolamiol, D., Pazhouhandeh, M., Marrocco, K., Genschik, P., and Ziegler-Graff, V. (2007). The polerovirus F box protein P0 targets argonaute1 to suppress RNA silencing. *Curr. Biol.* **17:**1615–1621.

Bouche, N., Lauressergues, D., Gasciolli, V., and Vaucheret, H. (2006). An antagonistic function for *Arabidopsis* DCL2 in development and a new function for DCL4 in generating viral siRNAs. *EMBO J.* **25:**3347–3356.

Brigneti, G., Voinnet, O., Li, W. X., Ji, L. H., Ding, S. W., and Baulcombe, D. C. (1998). Viral pathogenicity determinants are suppressors of transgene silencing in *Nicotiana benthamiana. EMBO J.* **17:**6739–6746.

Brodersen, P., Sakvarelidze-Achard, L., Bruun-Rasmussen, M., Dunoyer, P., Yamamoto, Y. Y., Sieburth, L., and Voinnet, O. (2008). Widespread translational inhibition by plant miRNAs and siRNAs. *Science* **320:**1185–1190.

Bucher, E., Sijen, T., De Haan, P., Goldbach, R., and Prins, M. (2003). Negative-strand tospoviruses and tenuiviruses carry a gene for a suppressor of gene silencing at analogous genomic positions. *J. Virol.* **77:**1329–1336.

Canizares, M. C., Taylor, K. M., and Lomonossoff, G. P. (2004). Surface-exposed C-terminal amino acids of the small coat protein of Cowpea mosaic virus are required for suppression of silencing. *J. Gen. Virol.* **85:**3431–3435.

Canizares, M. C., Navas-Castillo, J., and Moriones, E. (2008). Multiple suppressors of RNA silencing encoded by both genomic RNAs of the crinivirus, Tomato chlorosis virus. *Virology* **379:**168–174.

Cao, X., Zhou, P., Zhang, X., Zhu, S., Zhong, X., Xiao, Q., Ding, B., and Li, Y. (2005). Identification of an RNA silencing suppressor from a plant double-stranded RNA virus. *J. Virol.* **79:**13018–13027.

Chao, J. A., Lee, J. H., Chapados, B. R., Debler, E. W., Schneemann, A., and Williamson, J. R. (2005). Dual modes of RNA-silencing suppression by Flock House virus protein B2. *Nat. Struct. Mol. Biol.* **12:**952–957.

Chapman, E. J., Prokhnevsky, A. I., Gopinath, K., Dolja, V. V., and Carrington, J. C. (2004). Viral RNA silencing suppressors inhibit the microRNA pathway at an intermediate step. *Genes Dev.* **18:**1179–1186.

Chellappan, P., Vanitharani, R., and Fauquet, C. M. (2005). MicroRNA-binding viral protein interferes with *Arabidopsis* development. *Proc. Natl. Acad. Sci. U. S. A.* **102:**10381–10386.

Chen, J., Li, W. X., Xie, D., Peng, J. R., and Ding, S. W. (2004). Viral virulence protein suppresses RNA silencing-mediated defense but upregulates the role of microrna in host gene expression. *Plant Cell* **16:**1302–1313.

Chen, H. Y., Yang, J., Lin, C., and Yuan, Y. A. (2008). Structural basis for RNA-silencing suppression by Tomato aspermy virus protein 2b. *EMBO Rep.* **9:**754–760.

Chiba, M., Reed, J. C., Prokhnevsky, A. I., Chapman, E. J., Mawassi, M., Koonin, E. V., Carrington, J. C., and Dolja, V. V. (2006). Diverse suppressors of RNA silencing enhance agroinfection by a viral replicon. *Virology* **346:**7–14.

Cuellar, W. J., Tairo, F., Kreuze, J. F., and Valkonen, J. P. (2008). Analysis of gene content in sweet potato chlorotic stunt virus RNA1 reveals the presence of the p22 RNA silencing suppressor in only a few isolates: implications for viral evolution and synergism. *J. Gen. Virol.* **89:**573–582.

Cuellar, W. J., Kreuze, J. F., Rajamaki, M. L., Cruzado, K. R., Untiveros, M., and Valkonen, J. P. (2009). Elimination of antiviral defense by viral RNase III. *Proc. Natl. Acad. Sci. U. S. A.* **106:**10354–10358.

Csorba, T., Bovi, A., Dalmay, T., and Burgyan, J. (2007). The p122 subunit of Tobacco mosaic virus replicase is a potent silencing suppressor and compromises both siRNA and miRNA mediated pathways. *J. Virol.* **81:** 11768–11780.

Dalmay, T., Hamilton, A., Rudd, S., Angell, S., and Baulcombe, D. C. (2000). An RNA-dependent RNA polymerase gene in Arabidopsis is required for posttranscriptional gene silencing mediated by a transgene but not by a virus. *Cell* **101:**543–553.

Dalmay, T., Horsefield, R., Braunstein, T. H., and Baulcombe, D. C. (2001). SDE3 encodes an RNA helicase required for post-transcriptional gene silencing in Arabidopsis. *EMBO J.* **20:**2069–2078.

Deleris, A., Gallego-Bartolome, J., Bao, J., Kasschau, K. D., Carrington, J. C., and Voinnet, O. (2006). Hierarchical action and inhibition of plant Dicer-like proteins in antiviral defense. *Science* **313:**68–71.

Diaz-Pendon, J. A., and Ding, S. W. (2008). Direct and indirect roles of viral suppressors of RNA silencing in pathogenesis. *Annu. Rev. Phytopathol.* **46:** 303–326.

Diaz-Pendon, J. A., Li, F., Li, W. X., and Ding, S. W. (2007). Suppression of antiviral silencing by cucumber mosaic virus 2b protein in Arabidopsis is associated with drastically reduced accumulation of three classes of viral small interfering RNAs. *Plant Cell* **19:**2053–2063.

Ding, S. W., and Voinnet, O. (2007). Antiviral Immunity Directed by Small RNAs. *Cell* **130:**413–426.

Donaire, L., Barajas, D., Martinez-Garcia, B., Martinez-Priego, L., Pagan, I., and Llave, C. (2008). Structural and genetic requirements for the biogenesis of tobacco rattle virus-derived small interfering RNAs. *J. Virol.* **82:**5167–5177.

Donaire, L., Wang, Y., Gonzalez-Ibeas, D., Mayer, K. F., Aranda, M. A., and Llave, C. (2009). Deep-sequencing of plant viral small RNAs reveals effective and widespread targeting of viral genomes. *Virology* **392:**203–214.

Dougherty, W. G., Lindbo, J. A., Smith, H. A., Parks, T. D., Swaney, S., and Proebsting, W. M. (1994). RNA-mediated virus resistance in transgenic plants: exploitation of a cellular pathway possibly involved in RNA degradation. *Mol. Plant-Microbe Interact.* **7:**544–552.

Dunoyer, P., Pfeffer, S., Fritsch, C., Hemmer, O., Voinnet, O., and Richards, K. E. (2002). Identification, subcellular localization and some properties of a cysteine-rich suppressor of gene silencing encoded by peanut clump virus. *Plant J.* **29:**555–567.

Dunoyer, P., Lecellier, C. H., Parizotto, E. A., Himber, C., and Voinnet, O. (2004). Probing the microRNA and small interfering RNA pathways with virus-encoded suppressors of RNA silencing. *Plant Cell* **16:**1235–1250.

Dunoyer, P., Himber, C., and Voinnet, O. (2005). DICER-LIKE 4 is required for RNA interference and produces the 21-nucleotide small interfering RNA component of the plant cell-to-cell silencing signal. *Nat. Genet.* **37:**1356–1360.

Eamens, A., Vaistij, F. E., and Jones, L. (2008). NRPD1a and NRPD1b are required to maintain post-transcriptional RNA silencing and RNA-directed DNA methylation in Arabidopsis. *Plant J.* **55:**596–606.

Ebhardt, H. A., Thi, E. P., Wang, M. B., and Unrau, P. J. (2005). Extensive 3′ modification of plant small RNAs is modulated by helper component-proteinase expression. *Proc. Natl. Acad. Sci. U. S. A.* **102:**13398–13403.

El-Shami, M., Pontier, D., Lahmy, S., Braun, L., Picart, C., Vega, D., Hakimi, M. A., Jacobsen, S. E., Cooke, R., and Lagrange, T. (2007). Reiterated WG/GW motifs form functionally and evolutionarily conserved ARGONAUTE-binding platforms in RNAi-related components. *Genes Dev.* **21:**2539–2544.

Eulalio, A., Huntzinger, E., and Izaurralde, E. (2008). GW182 interaction with Argonaute is essential for miRNA-mediated translational repression and mRNA decay. *Nat. Struct. Mol. Biol.* **15:**346–353.

Fang, Y., and Spector, D. L. (2007). Identification of Nuclear Dicing Bodies Containing Proteins for MicroRNA Biogenesis in Living Arabidopsis Plants. *Curr. Biol.* **17:**818–823.

Flores, R., Hernandez, C., Martinez de Alba, A. E., Daros, J. A., and Di Serio, F. (2005). Viroids and viroid-host interactions. *Annu. Rev. Phytopathol.* **43:**117–139.

Fukunaga, R., and Doudna, J. A. (2009). dsRNA with 5′ overhangs contributes to endogenous and antiviral RNA silencing pathways in plants. *EMBO J.* **28:**545–555.

Genoves, A., Navarro, J. A., and Pallas, V. (2006). Functional analysis of the five melon necrotic spot virus genome-encoded proteins. *J. Gen. Virol.* **87:** 2371–2380.

Glick, E., Zrachya, A., Levy, Y., Mett, A., Gidoni, D., Belausov, E., Citovsky, V., and Gafni, Y. (2008). Interaction with host SGS3 is required for suppression of RNA silencing by tomato yellow leaf curl virus V2 protein. *Proc. Natl. Acad. Sci. U. S. A.* **105:**157–161.

Gomez, G., Martinez, G., and Pallas, V. (2009). Interplay between viroid-induced pathogenesis and RNA silencing pathways. *Trends Plant Sci.* **14:**264–269.

Goto, K., Kobori, T., Kosaka, Y., Natsuaki, T., and Masuta, C. (2007). Characterization of silencing suppressor 2b of cucumber mosaic virus based on examination of its small RNA-binding abilities. *Plant Cell Physiol.* **48:**1050–1060.

Guo, H. S., and Ding, S. W. (2002). A viral protein inhibits the long range signaling activity of the gene silencing signal. *EMBO J.* **21:**398–407.

Haas, G., Azevedo, J., Moissiard, G., Geldreich, A., Himber, C., Bureau, M., Fukuhara, T., Keller, M., and Voinnet, O. (2008). Nuclear import of CaMV P6 is required for infection and suppression of the RNA silencing factor DRB4. *EMBO J.* **27:**2102–2112.

Hamilton, A. J., and Baulcombe, D. C. (1999). A species of small antisense RNA in posttranscriptional gene silencing in plants. *Science* **286:**950–952.

Hamilton, A., Voinnet, O., Chappell, L., and Baulcombe, D. (2002). Two classes of short interfering RNA in RNA silencing. *EMBO J.* **21:**4671–4679.

Hammond, S. M., Bernstein, E., Beach, D., and Hannon, G. J. (2000). An RNA-directed nuclease mediates post-transcriptional gene silencing in Drosophila cells. *Nature* **404:**293–296.

Hannon, G. J., and Conklin, D. S. (2004). RNA interference by short hairpin RNAs expressed in vertebrate cells. *Methods Mol. Biol.* **257:**255–266.

Harries, P. A., Palanichelvam, K., Bhat, S., and Nelson, R. S. (2008). Tobacco mosaic virus 126-kDa protein increases the susceptibility of *Nicotiana tabacum* to other viruses and its dosage affects virus-induced gene silencing. *Mol. Plant-Microbe Interact.* **21:**1539–1548.
He, X. J., Hsu, Y. F., Zhu, S., Wierzbicki, A. T., Pontes, O., Pikaard, C. S., Liu, H. L., Wang, C. S., Jin, H., and Zhu, J. K. (2009). An effector of RNA-directed DNA methylation in arabidopsis is an ARGONAUTE 4- and RNA-binding protein. *Cell* **137:**498–508.
Hemmes, H., Lakatos, L., Goldbach, R., Burgyan, J., and Prins, M. (2007). The NS3 protein of Rice hoja blanca tenuivirus suppresses RNA silencing in plant and insect hosts by efficiently binding both siRNAs and miRNAs. *RNA* **13:** 1079–1089.
Hiraguri, A., Itoh, R., Kondo, N., Nomura, Y., Aizawa, D., Murai, Y., Koiwa, H., Seki, M., Shinozaki, K., and Fukuhara, T. (2005). Specific interactions between Dicer-like proteins and HYL1/DRB-family dsRNA-binding proteins in Arabidopsis thaliana. *Plant Mol. Biol.* **57:**173–188.
Ho, T., Pallett, D., Rusholme, R., Dalmay, T., and Wang, H. (2006). A simplified method for cloning of short interfering RNAs from Brassica juncea infected with Turnip mosaic potyvirus and Turnip crinkle carmovirus. *J. Virol. Methods* **136:**217–223.
Howell, M. D., Fahlgren, N., Chapman, E. J., Cumbie, J. S., Sullivan, C. M., Givan, S. A., Kasschau, K. D., and Carrington, J. C. (2007). Genome-wide analysis of the RNA-DEPENDENT RNA POLYMERASE6/DICER-LIKE4 pathway in Arabidopsis reveals dependency on miRNA- and tasiRNA-directed targeting. *Plant Cell* **19:**926–942.
Hull, R. (2002). *Matthews' Plant Virology.* Fourth edn. Academic Press, San Diego, California, USA.
Hunter, C., Willmann, M. R., Wu, G., Yoshikawa, M., de la Luz Gutierrez-Nava, M., and Poethig, S. R. (2006). Trans-acting siRNA-mediated repression of ETTIN and ARF4 regulates heteroblasty in Arabidopsis. *Development* **133:**2973–2981.
Itaya, A., Zhong, X., Bundschuh, R., Qi, Y., Wang, Y., Takeda, R., Harris, A. R., Molina, C., Nelson, R. S., and Ding, B. (2007). A structured viroid RNA serves as a substrate for dicer-like cleavage to produce biologically active small RNAs but is resistant to RNA-induced silencing complex-mediated degradation. *J. Virol.* **81:**2980–2994.
Kasschau, K. D., and Carrington, J. C. (1998). A counterdefensive strategy of plant viruses: suppression of posttranscriptional gene silencing. *Cell* **95:**461–470.
Kasschau, K. D., Xie, Z., Allen, E., Llave, C., Chapman, E. J., Krizan, K. A., and Carrington, J. C. (2003). P1/HC-Pro, a viral suppressor of RNA silencing, interferes with Arabidopsis development and miRNA function. *Dev. Cell* **4:**205–217.
Kataya, A. R., Suliman, M. N., Kalantidis, K., and Livieratos, I. C. (2009). Cucurbit yellow stunting disorder virus p25 is a suppressor of post-transcriptional gene silencing. *Virus Res.* **145:**48–53.
Kim, V. N. (2005). Small RNAs: classification, biogenesis, and function. *Mol. Cells* **19:**1–15.

Kreuze, J. F., Savenkov, E. I., Cuellar, W., Li, X., and Valkonen, J. P. (2005). Viral class 1 RNase III involved in suppression of RNA silencing. *J. Virol.* **79:** 7227–7238.

Kubota, K., Tsuda, S., Tamai, A., and Meshi, T. (2003). Tomato mosaic virus replication protein suppresses virus-targeted posttranscriptional gene silencing. *J. Virol.* **77:**11016–11026.

Kumakura, N., Takeda, A., Fujioka, Y., Motose, H., Takano, R., and Watanabe, Y. (2009). SGS3 and RDR6 interact and colocalize in cytoplasmic SGS3/RDR6-bodies. *FEBS Lett.* **583:**1261–1266.

Lakatos, L., Csorba, T., Pantaleo, V., Chapman, E. J., Carrington, J. C., Liu, Y. P., Dolja, V. V., Calvino, L. F., Lopez-Moya, J. J., and Burgyan, J. (2006). Small RNA binding is a common strategy to suppress RNA silencing by several viral suppressors. *EMBO J.* **25:**2768–2780.

Lanet, E., Delannoy, E., Sormani, R., Floris, M., Brodersen, P., Crete, P., Voinnet, O., and Robaglia, C. (2009). Biochemical Evidence for Translational Repression by Arabidopsis MicroRNAs. *Plant Cell* **21:**1762–1768.

Lewsey, M., Robertson, F. C., Canto, T., Palukaitis, P., and Carr, J. P. (2007). Selective targeting of miRNA-regulated plant development by a viral counter-silencing protein. *Plant J.* **50:**240–252.

Li, J., Yang, Z., Yu, B., Liu, J., and Chen, X. (2005). Methylation protects miRNAs and siRNAs from a 3′-end uridylation activity in Arabidopsis. *Curr. Biol.* **15:**1501–1507.

Lin, S. S., Wu, H. W., Elena, S. F., Chen, K. C., Niu, Q. W., Yeh, S. D., Chen, C. C., and Chua, N. H. (2009). Molecular evolution of a viral non-coding sequence under the selective pressure of a miRNA-mediated silencing. *PLoS Pathog.* **5:**e1000312.

Lindbo, J. A., and Dougherty, W. G. (1992). Pathogen-derived resistance to a potyvirus: immune and resistant phenotypes in transgenic tobacco expressing altered forms of a potyvirus coat protein nucleotide sequence. *Mol. Plant-Microbe Interact.* **5:**144–153.

Lindbo, J. A., Silva-Rosales, L., Proebsting, W. M., and Dougherty, W. G. (1993). Induction of a highly specific antiviral state in transgenic plants: Implications for regulation of gene expression and virus resistance. *Plant Cell* **5:**1749–1759.

Liu, J., Carmell, M. A., Rivas, F. V., Marsden, C. G., Thomson, J. M., Song, J. J., Hammond, S. M., Joshua-Tor, L., and Hannon, G. J. (2004). Argonaute2 is the catalytic engine of mammalian RNAi. *Science* **305:**1437–1441.

Liu, L., Grainger, J., Canizares, M. C., Angell, S. M., and Lomonossoff, G. P. (2004). Cowpea mosaic virus RNA-1 acts as an amplicon whose effects can be counteracted by a RNA-2-encoded suppressor of silencing. *Virology* **323:**37–48.

Liu, J., Rivas, F. V., Wohlschlegel, J., Yates, J. R. 3rd, Parker, R., and Hannon, G. J. (2005). A role for the P-body component GW182 in microRNA function. *Nat. Cell Biol.* **7:**1261–1266.

Love, A. J., Laird, J., Holt, J., Hamilton, A. J., Sadanandom, A., and Milner, J. J. (2007). Cauliflower mosaic virus protein P6 is a suppressor of RNA silencing. *J. Gen. Virol.* **88:**3439–3444.

Lozsa, R., Csorba, T., Lakatos, L., and Burgyan, J. (2008). Inhibition of 3′ modification of small RNAs in virus-infected plants require spatial and temporal co-expression of small RNAs and viral silencing-suppressor proteins. *Nucleic Acids Res.* **36:**4099–4107.

Lu, R., Folimonov, A., Shintaku, M., Li, W. X., Falk, B. W., Dawson, W. O., and Ding, S. W. (2004). Three distinct suppressors of RNA silencing encoded by a 20-kb viral RNA genome. *Proc. Natl. Acad. Sci. U. S. A.* **101:**15742–15747.

Ma, J. B., Ye, K., and Patel, D. J. (2004). Structural basis for overhang-specific small interfering RNA recognition by the PAZ domain. *Nature* **429:** 318–322.

Mangwende, T., Wang, M. L., Borth, W., Hu, J., Moore, P. H., Mirkov, T. E., and Albert, H. H. (2009). The P0 gene of Sugarcane yellow leaf virus encodes an RNA silencing suppressor with unique activities. *Virology* **384:**38–50.

Martinez-Priego, L., Donaire, L., Barajas, D., and Llave, C. (2008). Silencing suppressor activity of the Tobacco rattle virus-encoded 16-kDa protein and interference with endogenous small RNA-guided regulatory pathways. *Virology* **376:**346–356.

Matzke, M. A., and Birchler, J. A. (2005). RNAi-mediated pathways in the nucleus. *Nat. Rev. Genet.* **6:**24–35.

Mayers, C. N., Palukaitis, P., and Carr, J. P. (2000). Subcellular distribution analysis of the cucumber mosaic virus 2b protein. *J. Gen. Virol.* **81:**219–226.

Mayo, M. A., and Ziegler-Graff, V. (1996). Molecular biology of luteoviruses. *Adv. Virus Res.* **46:**413–460.

Mbanzibwa, D. R., Tian, Y., Mukasa, S. B., and Valkonen, J. P. (2009). Cassava brown streak virus (Potyviridae) encodes a putative Maf/HAM1 pyrophosphatase implicated in reduction of mutations and a P1 proteinase that suppresses RNA silencing but contains no HC-Pro. *J. Virol.* **83:**6934–6940.

Meister, G., and Tuschl, T. (2004). Mechanisms of gene silencing by double-stranded RNA. *Nature* **431:**343–349.

Meng, C., Chen, J., Ding, S. W., Peng, J., and Wong, S. M. (2008). Hibiscus chlorotic ringspot virus coat protein inhibits trans-acting small interfering RNA biogenesis in Arabidopsis. *J. Gen. Virol.* **89:**2349–2358.

Merai, Z., Kerenyi, Z., Molnar, A., Barta, E., Valoczi, A., Bisztray, G., Havelda, Z., Burgyan, J., and Silhavy, D. (2005). Aureusvirus P14 is an efficient RNA silencing suppressor that binds double-stranded RNAs without size specificity. *J. Virol.* **79:**7217–7226.

Merai, Z., Kerenyi, Z., Kertesz, S., Magna, M., Lakatos, L., and Silhavy, D. (2006). Double-stranded RNA binding may be a general plant RNA viral strategy to suppress RNA silencing. *J. Virol.* **80:**5747–5756.

Mi, S., Cai, T., Hu, Y., Chen, Y., Hodges, E., Ni, F., Wu, L., Li, S., Zhou, H., Long, C., Chen, S., Hannon, G. J., and Qi, Y. (2008). Sorting of small RNAs into Arabidopsis argonaute complexes is directed by the 5′ terminal nucleotide. *Cell* **133:**116–127.

Moissiard, G., and Voinnet, O. (2006). RNA silencing of host transcripts by cauliflower mosaic virus requires coordinated action of the four Arabidopsis Dicer-like proteins. *Proc. Natl. Acad. Sci. U. S. A.* **103:**19593–19598.

Molnar, A., Csorba, T., Lakatos, L., Varallyay, E., Lacomme, C., and Burgyan, J. (2005). Plant virus-derived small interfering RNAs originate

predominantly from highly structured single-stranded viral RNAs. *J. Virol.* **79:** 7812–7818.
Morel, J. B., Godon, C., Mourrain, P., Beclin, C., Boutet, S., Feuerbach, F., Proux, F., and Vaucheret, H. (2002). Fertile hypomorphic ARGONAUTE (ago1) mutants impaired in post-transcriptional gene silencing and virus resistance. *Plant Cell* **14:**629–639.
Nakazawa, Y., Hiraguri, A., Moriyama, H., and Fukuhara, T. (2007). The dsRNA-binding protein DRB4 interacts with the Dicer-like protein DCL4 in vivo and functions in the trans-acting siRNA pathway. *Plant. Mol. Biol.* **63:**777–785.
Ohara, T., Sakaguchi, Y., Suzuki, T., Ueda, H., and Miyauchi, K. (2007). The 3′ termini of mouse Piwi-interacting RNAs are 2′-O-methylated. *Nat. Struct. Mol. Biol.* **14:**349–350.
Omarov, R. T., Ciomperlik, J. J., and Scholthof, H. B. (2007). RNAi-associated ssRNA-specific ribonucleases in Tombusvirus P19 mutant-infected plants and evidence for a discrete siRNA-containing effector complex. *Proc. Natl. Acad. Sci. U. S. A.* **104:**1714–1719.
Pantaleo, V., and Burgyan, J. (2008). Cymbidium ringspot virus harnesses RNA silencing to control the accumulation of virus parasite satellite RNA. *J. Virol.* **82:**11851–11858.
Pantaleo, V., Szittya, G., and Burgyan, J. (2007). Molecular Bases of Viral RNA Targeting by Viral Small Interfering RNA-Programmed RISC. *J. Virol.* **81:** 3797–3806.
Pazhouhandeh, M., Dieterle, M., Marrocco, K., Lechner, E., Berry, B., Brault, V., Hemmer, O., Kretsch, T., Richards, K. E., Genschik, P., and Ziegler-Graff, V. (2006). F-box-like domain in the polerovirus protein P0 is required for silencing suppressor function. *Proc. Natl. Acad. Sci. U. S. A.* **103:**1994–1999.
Pfeffer, S., Dunoyer, P., Heim, F., Richards, K. E., Jonard, G., and Ziegler-Graff, V. (2002). P0 of beet Western yellows virus is a suppressor of posttranscriptional gene silencing. *J. Virol.* **76:**6815–6824.
Pham, J. W., Pellino, J. L., Lee, Y. S., Carthew, R. W., and Sontheimer, E. J. (2004). A Dicer-2-dependent 80s complex cleaves targeted mRNAs during RNAi in *Drosophila*. *Cell* **117:**83–94.
Plasterk, R. H. (2002). RNA silencing: the genome's immune system. *Science* **296:**1263–1265.
Powers, J. G., Sit, T. L., Heinsohn, C., George, C. G., Kim, K. H., and Lommel, S. A. (2008). The Red clover necrotic mosaic virus RNA-2 encoded movement protein is a second suppressor of RNA silencing. *Virology* **381:**277–286.
Qi, Y., Denli, A. M., and Hannon, G. J. (2005). Biochemical specialization within Arabidopsis RNA silencing pathways. *Mol. Cell.* **19:**421–428.
Qi, X., Bao, F. S., and Xie, Z. (2009). Small RNA deep sequencing reveals role for Arabidopsis thaliana RNA-dependent RNA polymerases in viral siRNA biogenesis. *PLoS ONE* **4:**e4971.
Qiu, W., and Scholthof, K. B. (2004). Satellite panicum mosaic virus capsid protein elicits symptoms on a nonhost plant and interferes with a suppressor of virus-induced gene silencing. *Mol. Plant-Microbe Interact.* **17:**263–271.

Qu, F., Ye, X., Hou, G., Sato, S., Clemente, T. E., and Morris, T. J. (2005). RDR6 has a broad-spectrum but temperature-dependent antiviral defense role in *Nicotiana benthamiana*. *J. Virol.* **79:**15209–15217.

Qu, F., Ye, X., and Morris, T. J. (2008). Arabidopsis DRB4, AGO1, AGO7, and RDR6 participate in a DCL4-initiated antiviral RNA silencing pathway negatively regulated by DCL1. *Proc. Natl. Acad. Sci. U. S. A.* **105:**14732–14737.

Rahim, M. D., Andika, I. B., Han, C., Kondo, H., and Tamada, T. (2007). RNA4-encoded p31 of beet necrotic yellow vein virus is involved in efficient vector transmission, symptom severity and silencing suppression in roots. *J. Gen. Virol.* **88:**1611–1619.

Ramachandran, V., and Chen, X. (2008). Degradation of microRNAs by a family of exoribonucleases in Arabidopsis. *Science* **321:**1490–1492.

Ratcliff, F., Harrison, B. D., and Baulcombe, D. C. (1997). A Similarity Between Viral Defense and Gene Silencing in Plants. *Science* **276:**1558–1560.

Reed, J. C., Kasschau, K. D., Prokhnevsky, A. I., Gopinath, K., Pogue, G. P., Carrington, J. C., and Dolja, V. V. (2003). Suppressor of RNA silencing encoded by Beet yellows virus. *Virology* **306:**203–209.

Rivas, F. V., Tolia, N. H., Song, J. J., Aragon, J. P., Liu, J., Hannon, G. J., and Joshua-Tor, L. (2005). Purified Argonaute2 and an siRNA form recombinant human RISC. *Nat. Struct. Mol. Biol.* **12:**340–349.

Saunders, K., Norman, A., Gucciardo, S., and Stanley, J. (2004). The DNA beta satellite component associated with ageratum yellow vein disease encodes an essential pathogenicity protein (betaC1). *Virology* **324:**37–47.

Schwach, F., Vaistij, F. E., Jones, L., and Baulcombe, D. C. (2005). An RNA-dependent RNA polymerase prevents meristem invasion by potato virus X and is required for the activity but not the production of a systemic silencing signal. *Plant Physiol.* **138:**1842–1852.

Shen, B., and Goodman, H. M. (2004). Uridine addition after microRNA-directed cleavage. *Science* **306:**997.

Silhavy, D., Molnar, A., Lucioli, A., Szittya, G., Hornyik, C., Tavazza, M., and Burgyan, J. (2002). A viral protein suppresses RNA silencing and binds silencing-generated, 21- to 25-nucleotide double-stranded RNAs. *EMBO J.* **21:**3070–3080.

Simon, A. E., Roossinck, M. J., and Havelda, Z. (2004). Plant virus satellite and defective interfering RNAs: new paradigms for a new century. *Annu. Rev. Phytopathol.* **42:**415–437.

Simon-Mateo, C., and Garcia, J. A. (2006). MicroRNA-guided processing impairs Plum pox virus replication, but the virus readily evolves to escape this silencing mechanism. *J. Virol.* **80:**2429–2436.

Sire, C., Bangratz-Reyser, M., Fargette, D., and Brugidou, C. (2008). Genetic diversity and silencing suppression effects of Rice yellow mottle virus and the P1 protein. *Virol. J.* **5:**55.

Soards, A. J., Murphy, A. M., Palukaitis, P., and Carr, J. P. (2002). Virulence and differential local and systemic spread of cucumber mosaic virus in tobacco are affected by the CMV 2b protein. *Mol. Plant-Microbe Interact.* **15:**647–653.

Song, J. J., Smith, S. K., Hannon, G. J., and Joshua-Tor, L. (2004). Crystal structure of Argonaute and its implications for RISC slicer activity. *Science* **305:**1434–1437.

Szittya, G., Molnar, A., Silhavy, D., Hornyik, C., and Burgyan, J. (2002). Short defective interfering RNAs of tombusviruses are not targeted but trigger post-transcriptional gene silencing against their helper virus. *Plant Cell* **14:**359–372.

Takeda, A., Sugiyama, K., Nagano, H., Mori, M., Kaido, M., Mise, K., Tsuda, S., and Okuno, T. (2002). Identification of a novel RNA silencing suppressor, NSs protein of Tomato spotted wilt virus. *FEBS Lett.* **532:**75–79.

Takeda, A., Tsukuda, M., Mizumoto, H., Okamoto, K., Kaido, M., Mise, K., and Okuno, T. (2005). A plant RNA virus suppresses RNA silencing through viral RNA replication. *EMBO J.* **24:**3147–3157.

Takeda, A., Iwasaki, S., Watanabe, T., Utsumi, M., and Watanabe, Y. (2008). The mechanism selecting the guide strand from small RNA duplexes is different among argonaute proteins. *Plant Cell Physiol.* **49:**493–500.

Te, J., Melcher, U., Howard, A., and Verchot-Lubicz, J. (2005). Soilborne wheat mosaic virus (SBWMV) 19K protein belongs to a class of cysteine rich proteins that suppress RNA silencing. *Virol. J.* **2:**18.

Thomas, C. L., Leh, V., Lederer, C., and Maule, A. J. (2003). Turnip crinkle virus coat protein mediates suppression of RNA silencing in *Nicotiana benthamiana*. *Virology* **306:**33–41.

Tolia, N. H., and Joshua-Tor, L. (2007). Slicer and the argonautes. *Nat. Chem. Biol.* **3:**36–43.

Tomari, Y., and Zamore, P. D. (2005). Perspective: machines for RNAi. *Genes. Dev.* **19:**517–529.

Trinks, D., Rajeswaran, R., Shivaprasad, P. V., Akbergenov, R., Oakeley, E. J., Veluthambi, K., Hohn, T., and Pooggin, M. M. (2005). Suppression of RNA silencing by a geminivirus nuclear protein, AC2, correlates with transactivation of host genes. *J. Virol.* **79:**2517–2527.

Vaistij, F. E., and Jones, L. (2009). Compromised virus-induced gene silencing in RDR6-deficient plants. *Plant Physiol.* **149:**1399–1407.

Valli, A., Dujovny, G., and Garcia, J. A. (2008). Protease activity, self interaction, and small interfering RNA binding of the silencing suppressor p1b from cucumber vein yellowing ipomovirus. *J. Virol.* **82:**974–986.

Vanitharani, R., Chellappan, P., Pita, J. S., and Fauquet, C. M. (2004). Differential roles of AC2 and AC4 of cassava geminiviruses in mediating synergism and suppression of posttranscriptional gene silencing. *J. Virol.* **78:**9487–9498.

Vargason, J., Szittya, G., Burgyan, J., and Hall, T. M. (2003). Size selective recognition of siRNA by an RNA silencing suppressor. *Cell* **115:**799–811.

Vaucheret, H. (2006). Post-transcriptional small RNA pathways in plants: mechanisms and regulations. *Genes Dev.* **20:**759–771.

Verdel, A., Jia, S., Gerber, S., Sugiyama, T., Gygi, S., Grewal, S. I., and Moazed, D. (2004). RNAi-mediated targeting of heterochromatin by the RITS complex. *Science* **303:**672–676.

Vogler, H., Akbergenov, R., Shivaprasad, P. V., Dang, V., Fasler, M., Kwon, M. O., Zhanybekova, S., Hohn, T., and Heinlein, M. (2007). Modification of small

RNAs associated with suppression of RNA silencing by tobamovirus replicase protein. *J. Virol.* **81:**10379–10388.

Voinnet, O. (2002). RNA silencing: small RNAs as ubiquitous regulators of gene expression. *Curr. Opin. Plant Biol.* **5:**444.

Voinnet, O. (2005). Induction and suppression of RNA silencing: insights from viral infections. *Nat. Rev. Genet.* **6:**206–220.

Voinnet, O., Pinto, Y. M., and Baulcombe, D. C. (1999). Suppression of gene silencing: a general strategy used by diverse DNA and RNA viruses of plants. *Proc. Natl. Acad. Sci. U. S. A.* **96:**14147–14152.

Voinnet, O., Lederer, C., and Baulcombe, D. C. (2000). A viral movement protein prevents spread of the gene silencing signal in *Nicotiana benthamiana*. *Cell* **103:**157–167.

Wang, H., Hao, L., Shung, C. Y., Sunter, G., and Bisaro, D. M. (2003). Adenosine kinase is inactivated by geminivirus AL2 and L2 proteins. *Plant Cell* **15:**3020–3032.

Wang, H., Buckley, K. J., Yang, X., Buchmann, R. C., and Bisaro, D. M. (2005). Adenosine kinase inhibition and suppression of RNA silencing by geminivirus AL2 and L2 proteins. *J. Virol.* **79:**7410–7418.

Wassenegger, M., and Krczal, G. (2006). Nomenclature and functions of RNA-directed RNA polymerases. *Trends Plant Sci.* **11:**142–151.

Xie, Z., Fan, B., Chen, C., and Chen, Z. (2001). An important role of an inducible RNA-dependent RNA polymerase in plant antiviral defense. *Proc. Natl. Acad. Sci. U. S. A.* **98:**6516–6521.

Xiong, R., Wu, J., Zhou, Y., and Zhou, X. (2009). Characterization and subcellular localization of an RNA silencing suppressor encoded by Rice stripe tenuivirus. *Virology* **387:**29–40.

Yaegashi, H., Takahashi, T., Isogai, M., Kobori, T., Ohki, S., and Yoshikawa, N. (2007). Apple chlorotic leaf spot virus 50 kDa movement protein acts as a suppressor of systemic silencing without interfering with local silencing in *Nicotiana benthamiana*. *J. Gen. Virol.* **88:**316–324.

Yaegashi, H., Tamura, A., Isogai, M., and Yoshikawa, N. (2008). Inhibition of long-distance movement of RNA silencing signals in *Nicotiana benthamiana* by Apple chlorotic leaf spot virus 50 kDa movement protein. *Virology* **382:**199–206.

Yang, S. J., Carter, S. A., Cole, A. B., Cheng, N. H., and Nelson, R. S. (2004). A natural variant of a host RNA-dependent RNA polymerase is associated with increased susceptibility to viruses by *Nicotiana benthamiana*. *Proc. Natl. Acad. Sci. U. S. A.* **101:**6297–6302.

Yang, J. H., Seo, H. H., Han, S. J., Yoon, E. K., Yang, M. S., and Lee, W. S. (2008). Phytohormone abscisic acid control RNA-dependent RNA polymerase 6 gene expression and post-transcriptional gene silencing in rice cells. *Nucleic Acids Res.* **36:**1220–1226.

Ye, K., Malinina, L., and Patel, D. J. (2003). Recognition of small interfering RNA by a viral suppressor of RNA silencing. *Nature* **426:**874–878.

Yelina, N. E., Savenkov, E. I., Solovyev, A. G., Morozov, S. Y., and Valkonen, J. P. (2002). Long-distance movement, virulence, and RNA silencing suppression controlled by a single protein in hordei- and potyviruses: complementary functions between virus families. *J. Virol.* **76:**12981–12991.

Yu, D., Fan, B., MacFarlane, S. A., and Chen, Z. (2003). Analysis of the involvement of an inducible Arabidopsis RNA-dependent RNA polymerase in antiviral defense. *Mol. Plant-Microbe Interact.* **16:**206–216.

Yu, B., Yang, Z., Li, J., Minakhina, S., Yang, M., Padgett, R. W., Steward, R., and Chen, X. (2005). Methylation as a crucial step in plant microRNA biogenesis. *Science* **307:**932–935.

Zamore, P. D. (2002). Ancient pathways programmed by small RNAs. *Science* **296:**1265–1269.

Zhang, X., Yuan, Y. R., Pei, Y., Lin, S. S., Tuschl, T., Patel, D. J., and Chua, N. H. (2006). Cucumber mosaic virus-encoded 2b suppressor inhibits Arabidopsis Argonaute1 cleavage activity to counter plant defense. *Genes Dev.* **20:** 3255–3268.

Zhou, Z., Dell'Orco, M., Saldarelli, P., Turturo, C., Minafra, A., and Martelli, G. P. (2006). Identification of an RNA-silencing suppressor in the genome of Grapevine virus A. *J. Gen. Virol.* **87:**2387–2395.

Zrachya, A., Glick, E., Levy, Y., Arazi, T., Citovsky, V., and Gafni, Y. (2007). Suppressor of RNA silencing encoded by Tomato yellow leaf curl virus-Israel. *Virology* **358:**159–165.

CHAPTER 3

Local Lesions and Induced Resistance☆

G. Loebenstein

Contents

Abstract The local lesion phenomenon is one of the most notable resistance mechanisms where virus after multiplying in several hundred cells around the point of entry, does not continue to

☆Parts of this chapter were adapted from: Loebenstein, G. and Akad, F. (2006). The local lesion response. *In* "Natural Resistance Mechanisms of Plants to Viruses" (G. Loebenstein and J.P. Carr, eds.) pp. 99-124. Springer, Dordrecht, The Netherlands.

Professor Emeritus, Agricultural Research Organization, Bet Dagan 50-250, Israel

Advances in Virus Research, Volume 75
ISSN: 0065-3527, DOI: 10.1016/S0065-3527(09)07503-4

spread and remains in a local infection. Several types of local lesions are known, *inter alia,* necrotic, chlorotic, and starch lesions. Cells inside the lesion generally contain much less virus than cells in a systemic infection. Cytopathic changes accompany the local lesion development. Proteases that may have properties similar to caspases, which promote programmed cell death (PCD) in animals, seem to participate in PCD during the hypersensitive response. Salicylic acid seems to be associated with the HR and may play a role in localizing the virus. The functions and properties of the *N* gene of *Nicotiana*, which was the first plant virus resistance gene to be isolated by transposon tagging, are discussed and compared with other plant genes for disease resistance. The Inhibitor of Virus Replication (IVR) associated with the local lesion response is mainly a tetratricopeptide repeat (TPR) protein. TPR motifs are also present in inducible interferons found in animal cells. Transformation of *N. tabacum* cv. Samsun nn, in which *Tobacco mosaic virus* (TMV) spreads systemically, with the NC330 gene sequence, encoding an IVR-like protein, resulted in a number of transgenic plant lines, expressing variable resistance to TMV and the fungal pathogen *Botrytis cinerea.* Transformation of tomato plants with the IVR gene became also partially resistant to *B. cinerea* (Loebenstein et al., in press). IVR-like compounds were found in the interspecific hybrid of *N. glutinosa x N. debneyi* that is highly resistant to TMV, and in the "green island" tissue of tobacco, cv. Xanthi-nc, infected with *Cucumber mosaic virus* (CMV).

Infection in one part of the plant often induces resistance in other non-invaded tissues. Local (LAR) or systemic (SAR) acquired resistance can be activated by viruses, bacterial, and fungal pathogens or other natural and synthetic compounds. Accumulation of salicylic acid accompanies the induction of resistance. Possible mechanisms are outlined. Synthetic compounds, as for example, acibenzolar-*S*-methyl (ASM) were developed for use in a novel strategy for crop protection through abiotic induction of SAR. For example, ASM protected cantaloupes against a fungal pathogen and CMV. Additional attempts to protect crops by inducing SAR are outlined and it is hoped that future research and its application will find its use in plant protection.

I. LOCAL LESIONS

Viruses that spread systemically in plants can cause economically important diseases. However, in several laboratory test or indicator plants the virus, after multiplying in several hundred cells around the point of entry, does not continue to spread and remains in a local infection.

Francis O. Holmes (1897–1990) who had a pioneering interest in virus mutants characterized the biological properties of *Tobacco mosaic virus* (TMV) variants. Holmes showed that strains (mutants) of TMV, in a spectrum of host plant species gave both a susceptible or resistant interaction, and the phenotype of the TMV mutants and the genetic effects of a host gene could be determined by observing the symptoms on plants. His discovery that TMV induces local lesions on certain *Nicotiana* sp. (Holmes, 1929) and the local lesion assay became a commonplace in plant virus laboratories till today.

Several types of local infections are known (Loebenstein *et al.*, 1982): (a) self-limiting necrotic local lesions such as those induced by TMV in *Datura stramonium*, where lesions reach their maximum size 3 days after inoculation; (b) chlorotic local lesions, such as *Potato virus Y* (PVY) induces in *Chenopodium amaranticolor*, where infected cells lose chlorophyll; some diseases as, *Passion fruit green spot virus, Coffee ringspot virus, and Citrus leprosis virus* which are transmitted by *Brevipalpus* mite species are characterized by localized lesions (chlorotic, green spots, or ringspots) on leaves, stems, and fruits (Kitajima *et al.*, 2003); (c) ring-like patterns or ringspots that remain localized, such as seen in *Tetragonia expansa* infected with *Tomato spotted wilt virus* (TSWV); (d) starch lesions, such as those triggered by TMV in cucumber cotyledons, where no symptoms are observed on the intact leaf, but when it is decolorized with ethanol and stained with iodine, lesions become apparent; (e) microlesions (with a mean size of 1.1×10^{-2} mm^2), such as the lesions induced by the U_2 strain of TMV on Pinto bean leaves; and (f) subliminal symptomless infections not detectable as starch lesions, as in TMV-infected cotton cotyledons, where virus content is 1/200,000 of that produced in a susceptible host (Cheo, 1970). The localized infection is an efficient mechanism whereby plants resist viruses, though most viral resistance genes are not associated with the hypersensitive response (HR), but affect virus multiplication or movement as a result of incompatible viral and host factors). The local lesion infection is one of the most notable resistance responses and has been used by breeders to obtain resistant cultivars of tobacco and sweet peppers against TMV (Kang *et al.*, 2005).

The zone around a TMV lesion on *NN*-genotype tobacco is also resistant to other strains of TMV (tomato aucuba virus and Holmes' ribgrass strain), *Tobacco necrosis virus* (TNV) and *Tomato ringspot virus* (ToRSV), but not to *Turnip mosaic virus* (TuMV). TNV also induced localized resistance to TMV (Ross, 1961a). This type of induced resistance was called *localized acquired resistance* (LAR). Apparently virus replication is inhibited in the zone around the lesion.

In *Nicotiana glutinosa* local lesion cells infected with TMV the number of virus particles per cell is about 10^3 (Milne, 1966). This is two to four orders of magnitude lower than in a comparable systemic infection, where the number of particles per cell is estimated to be between 10^5 and 6×10^7 (Harrison, 1955). TMV content (as measured by local lesion counts) in protoplasts of *N. tabacum* Samsun NN, where TMV induces local lesions, was about 1/4–1/10 of that in protoplasts of *N. tabacum* Samsun, where TMV spreads systemically in the intact plant (Loebenstein *et al.*, 1980). [That in isolated protoplasts from these two cultivars TMV multiplies to the same extent (Otsuki *et al.*, 1972) was due to the presence of 2,4-dichlorophenoxyacetic acid (2,4-D) in the protoplast incubation medium (Loebenstein *et al.*, 1980). 2,4-D has been reported to suppress localization and to enhance virus multiplication in local lesion intact hosts (Simons and Ross, 1965)]. These data indicated that localization is at least partly due to reduced multiplication in the cells of these hosts and not due to barrier substances, implicated in early research as possible factors preventing virus movement.

Most tobamoviruses induce local lesions in tobacco containing the *N* gene (see below). However, the tomato mosaic virus-OB strain overcomes the *N* gene-mediated HR (Tobias *et al.*, 1982). Padgett and Beachy (1993) showed that properties of the movement protein (see below) are not responsible for the resistance-breaking character of the OB strain.

The *N* gene–associated resistance ultimately affects both cell-to-cell movement and long-distance movement of TMV. In tobacco NN plants kept at temperatures above 28°C this restriction is inactive and TMV spreads throughout the plant in the same way as it spreads through nn-genotype tobacco. Restoring the temperature to below 28°C again allows activation of the *N* gene-mediated recognition of the virus, resulting in necrosis of all the tissues containing TMV and restricting further virus movement. However, movement of a TMV-based vector expressing green fluorescent protein (TMV-GFP) was still very limited in *NN*-genotype tobacco even when these plants were incubated at 33°C. This restriction of the spread of TMV-GFP was maintained in transgenic plants expressing a salicylic acid-degrading enzyme. In contrast, TMV-GFP moved efficiently in *NN*-genotype tobacco plants if they were transformed with a transgene derived from RNA1 of CMV, which encodes the CMV 1a protein (Canto and Palukaitis, 2002). These findings indicated a novel temperature-independent resistance to the movement of viruses, which operates via a pathway independent of salicylic acid but which can be subverted in some way by the CMV 1a protein (Canto and Palukaitis, 2002).

The *L* gene in Tabasco and *Capsicum chinense* peppers confers hypersensitivity to infection with TMV. Several alleles occur at a single locus and are partially dominant (Boukema, 1980). The L^2 gene is effective against all of the pepper-infecting tobamoviruses except *Pepper mild mottle virus* (PMMoV), whereas that conferred by the L^4 gene is effective against them all. The coat protein of tobamovirus acts as elicitor of both L^2 and L^4 gene-mediated resistance in *Capsicum* (Gilardi *et al.*, 2004). Two amino acid substitutions in the coat protein, Gln to Arg at position 46 and Gly to Lys at position 85, were responsible for overcoming the L^4 resistance gene (Genda *et al.*, 2007). Also a L^3 resistance-breaking field isolate of PMMoV, designated PMMoV-Is, had two amino acid changes in its coat protein, namely leucine to phenylalanine at position 13 (L13F) and glycine to valine at position 66 (G66V), as compared with PMMoV-J, which induces a resistance response in L^3-harboring *Capsicum* plants (Hamada *et al.*, 2007). A single dominant gene in beans controls local lesion formation by *Southern bean mosaic virus* (Holmes, 1954).

Cauliflower mosaic virus (CaMV) induces necrotic local lesions in tobacco and *Datura*, *Tomato bushy stunt virus* in tobacco, CMV in cowpea, *Potato virus X* (PVX) in potato cultivars carrying the *Nx* or *Nb* genes (Cockerham, 1955) and *Barley stripe mosaic virus* in *Chenopodium amaranticolor*.

In *Arabidopsis thaliana*, *Turnip crinkle virus* (TCV) produces an HR 2–3 days postinoculation in ecotype Dijon (Dempsey *et al.*, 1993; Simon *et al.*, 1992). Other ecotypes of *Arabidopsis* tested did not give an HR but allowed systemic spread of the virus (Li and Simon, 1990). A dominant gene, *HRT*, which confers an HR to TCV, has been identified and mapped in the Di-17 line of Dijon (Dempsey *et al.*, 1997). A coiled-coil, nucleotide-binding site (NBS)-leucine-rich repeat-type resistance gene, *RCY1*, confers resistance to a yellow strain of CMV(Y), in *Arabidopsis* ecotype C24. Resistance to CMV(Y) in C24 is accompanied by a HR that is characterized by the development of necrotic local lesions at the primary infection sites. Transforming ecotype Col-0, which is susceptible to CMV (Y) with *RCY1*completely suppressed systemic spread of CMV(Y). However, the virulent strain CMV (B2) spread and multiplied systemically in these transgenic lines similar to that in wild-type Col-0. Apparently constitutive accumulation of high levels of RCY1 protein appears to regulate the strength of *RCY1*-conferred resistance (Sekine *et al.*, 2008).

Some strains of TMV, such as the crucifer- and garlic-infecting TMV strain Cg, cause necrotic local lesions on Samsun nn tobacco and do not move systemically (Ehrenfeld *et al.*, 2005). Integrity of the coat proteins is important for triggering the HR-like response (Ehrenfeld *et al.*, 2008). The broad bean strain of TMV (TMV-B) infects *N. tabacum* White Burley systemically whereas the tomato strain of *Tomato mosaic virus* (ToMV-S1) induces necrotic local lesions and is restricted to inoculated leaves (Wu and Zhou, 2002).

Apparently differences in movement protein sequences influence the size of the local lesion, as shown using a chimaeric virus. When the movement protein of the TMV-B strain was exchanged with that of TomV-S1 resulted in small lesions. Also, C-terminal phosphorylation of the TMV movement protein abolishes its ability to promote virus spread (Waigmann *et al.*, 2000); and mutation of Thr to Asp mimicking Thr^{104} phosphorylation strongly inhibited cell-to-cell movement resulting in small lesions on *N. tabacum* Xanthi nc plants (Karger *et al.*, 2003). Some strains of *Tomato bushy stunt virus* elicit necrotic local lesions on *N. edwardsonii* plants. The p22 protein domain controls both viral cell-to-cell movement function and those that elicit necrotic lesions on selected resistant plants. Using site specific mutants Chu *et al.* (1999) showed that the p22 protein residues essential for movement are separable from those crucial for elicitation of the necrotic resistance response.

Extremely low frequency magnetic fields (ELF-MFs) applied to tobacco plants reacting to TMV with a hypersensitive response (HR) showed a decrease in lesion area and number (Trebbi *et al.*, 2007).

A. Cytopathic changes

In the necrotic lesion the cytopathic events leading to collapse and necrosis start with the induction of a reactive oxygen burst observed within seconds of the addition of exogenous TMV to the outside of tobacco (*Nicotiana tabacum* cv Samsun NN, EN, or nn) epidermal cells (Allan *et al.*, 2001). Changes in the chloroplast, consisting of swelling and distortion of the chlorophyll lamella and swelling of starch grains, are observed about 8 h after infection (Weintraub and Ragetli, 1964). This led to increased electrolyte leakage, even before local lesions became visible (Weststeijn, 1978), and, for example, the superoxidic radical monohydroascorbate increased markedly when TMV lesions on Xanthi-nc-tobacco leaves developed (Fodor *et al.*, 2001).

Incubation of NN tobacco plants to 32°C or higher inhibits the *N* gene-mediated HR. Transfer back to 20°C initiates the HR again. Using fluorescent-tagged TMV revealed membrane damage, which preceded visible cell collapse by more than 3 h, and was accompanied by a transient restriction of the xylem within infection sites on *N. edwardsonii.* Following cell collapse and the rapid desiccation of tissue undergoing the HR, isolated, infected cells were detected at the margin of necrotic lesions. These virus-infected cells were able to reinitiate infection on transfer to 32°C, however, if maintained at 20°C they eventually died. The results indicate that the TMV-induced HR is a two-phase process with an early stage culminating in rapid cell collapse and tissue desiccation followed by a more extended period during which the remaining infected cells are eliminated (Wright *et al.*, 2000).

This PCD was accompanied by accumulation of PR1, a protein that is induced during the HR (Linthorst, 1991). Okadaic acid (OA), an inhibitor of type 1 and type 2A serine/threonine protein phosphatases, can block both *N* gene-mediated HR and developmental PCD in plants (Dunigan and Madlener, 1995; Lacomme and Santa Cruz, 1999). It was suggested that the TMV-mediated lethal HR in plants requires reversible-protein phosphorylation in a signaling pathway that initiates the cell-death program; however, after entering the execution phase of the program, the process becomes irreversible (Lacomme and Santa Cruz, 1999).

There are some parallels between cell death in the plant HR and animal PCD. One similarity between plant and animal cell-death processes is the apparent role of protein phosphatase(s), which is required for developmental and pathogen Caspase-triggered PCD in plants (Dunigan and Madlener, 1995), and is also implicated in animal PCD where protein phosphatase 2A activity is specifically up-regulated by a cell-death-related protease (Morana *et al.*, 1996). A key event in animal PCD is the release of cytochrome *c* from mitochondria into the cytosol, initiating the final degradation phase of the cell-death program (Green and Reed, 1998). The ability of the GFP-TM fusion protein to target mitochondria suggests that mitochondrial targeting is also necessary for the plant response (Lacomme and Santa Cruz, 1999). Caspase-like proteases, known to suppress PCD in animals, seem to participate during HR. The p^{35} protein from baculovirus is a broad-range caspase inhibitor, and infection of p^{35} - expressing *N* tobacco plants with TMV disrupted *N*-mediated resistance, lead to systemic spreading of the virus (Pozo and Lam, 2003).

Seo *et al.* (2000) investigated genes involved in this hypersensitive reaction. They isolated the cDNA of tobacco DS9, the transcript of which decreases before the appearance of necrotic lesions. The DS9 gene encodes a chloroplastic homolog of bacterial FtsH protein, which serves to maintain quality control of some cytoplasmic and membrane proteins. A large quantity of DS9 protein was found in healthy leaves, whereas the quantity of DS9 protein in infected leaves decreased before the lesions appeared. In transgenic tobacco plants containing less and more DS9 protein than wild-type plants, the necrotic lesions induced by TMV were smaller and larger, respectively, than those on wild-type plants. These results suggest that a decrease in the level of DS9 protein in TMV-infected cells, resulting in a subsequent loss of function of the chloroplasts, accelerates the hypersensitive reaction.

HR-resistance and cell death can be physiologically, genetically and temporally uncoupled (Király and Király, 2006). Thus, for example, the plant gene *CCD1* selectively blocks cell death during the HR to CaMV infection (Cawly *et al.*, 2005). HR requires cell-to-cell contact and is not expressed in protoplasts. Protoplasts from plants carrying the *N* gene do not respond to TMV infection with cell death, though TMV multiplication in them is reduced markedly (Loebenstein *et al.*, 1980).

Actinomycin D or chloramphenicol when added up to 24 h after inoculation (but not later) markedly increased TMV replication in protoplasts of tobacco NN, while no increase was observed in protoplasts of tobacco nn (Gera *et al.*, 1983). This indicated that HR and inhibition of virus replication associated with the *N* gene are two different processes, and that this mechanism which presumably requires DNA-dependent RNA synthesis for its operation produces a substance that inhibits virus replication (see Section I.I.).

In leaves of *Nicotiana edwardsonii*, an interspecific hybrid derived from a cross between *N. glutinosa* and *N. clevelandii* CaMV strain W260 elicits an HR. *N. glutinosa* is resistant to W260, but responds with local chlorotic lesions rather than necrotic lesions. In contrast, *N. clevelandii* responds to W260 with systemic cell death. It was shown that the resistance and cell death that comprise the HR elicited by W260 could be uncoupled. The non-necrotic resistance response of *N. glutinosa* could be converted to HR when these plants were crossed with *N. clevelandii*. Also, cell death and resistance segregated independently in the F2 population of a cross between *N. edwardsonii* and *N. clevelandii*. Cole *et al.* (2001) concluded that the resistance of *N. edwardsonii* to W260 infection was conditioned by a gene derived from *N. glutinosa*, whereas a gene derived from *N. clevelandii* conditioned cell death.

In chlorotic local lesions, induced by TuMV in *C. quinoa*, disintegration of chloroplasts was not observed, although most palisade and spongy mesophyll cells contained chlorotic chloroplasts forming large aggregates of up to 20, instead of being uniformly distributed along the cell periphery (Kitajima and Costa, 1973).

In starch lesion hosts, such as those induced by TMV in cucumber cotyledons, swelling of chloroplasts are first observed 2–2.5 days after inoculation. When viewed 5 days after inoculation these chloroplasts contain large starch grains, but neither they nor the cells disintegrate or die. In the peripheral cells of a starch lesion the number of TMV particles was about 1/10 compared with those in the central part of the starch lesion, with no barriers or ultrastructural changes at the border of the lesion (Cohen and Loebenstein, 1975). In early studies it was observed that when cucumber cotyledons were treated with actinomycin D, chloramphenicol or UV irradiation 1 day after inoculation with TMV virus concentration increased markedly, indicating that during localization a substance is produced that reduces virus multiplication (Loebenstein *et al.*, 1969, 1970; Sela *et al.*, 1969).

The HR (e.g., as induced in Samsun NN tobacco infected with TMV U1 or TNV, or in Samsun with TNV) is accompanied by a sharp rise in the production and accumulation of the ethylene precursor, 1-aminocyclopropane-1-carboxylic acid (ACC), followed by rapid evolution of ethylene when the capacity to convert ACC to ethylene had increased. None of these changes occurred in systemically infected susceptible plants (de Laat and van Loon, 1983). Both the increase in ACC production and its accumulation were restricted to the cells surrounding the necrotic areas.

B. Pathogenesis-related proteins (PR-proteins)

Necrotic lesion formation is associated with the induction of a number of PR proteins (Van Loon and Van Kammen, 1970). Tobacco PR proteins consist of at least five families, each of which contains both acidic and basic isoforms (Van Loon *et al.*, 1994). There are 14 families of PR protein (PR1 through 14), PR-2 and PR-3 having β-1,3-glucanase and chitinase activities, respectively (Kauffmann *et al.*, 1987; Legrand *et al.*, 1987). These proteins have been studied extensively as they can be detected easily by gel-electrophoresis, but so far no evidence has been provided that they are active in localizing the virus. They are mainly induced in virus infections that do cause necrosis and may therefore be a host response to necrosis (or other stresses). However, PR-proteins were also seen in non-necrotic systemic infections by some viruses as CMV and PVY (Whitham *et al.*, 2003). No antiviral activity of any of the PR-proteins has so far been reported. Accumulation rates of PR proteins in *Capsicum chinense* L^3 plants infected with PMMoV did not correlate with maximal accumulation levels of viral RNA, thus indicating that PR protein expression may reflect the physiological status of the plant (Elvira *et al.*, 2008).

C. Compounds that induce resistance

Various compounds injected into the intercellular spaces of *NN*-genotype tobacco or *Datura stramonium* induced resistance to TMV, resulting in fewer and smaller lesions. Thus, yeast RNA (Gicherman and Loebenstein, 1968), poly I: poly C (Stein and Loebenstein, 1970), heat-killed cells of *Pseudomonas syringae* (Loebenstein and Lovrekovich, 1966), polyacrylic acid (Gianinazzi and Kassanis, 1974), several synthetic polyanions (Stahmann and Gothoskar, 1958; Stein and Loebenstein, 1970), root extracts of *Boerhaavia diffusa*—a glycoprotein, applied to the lower leaves (Verma and Awasthi, 1980; Verma *et al.*, 1979), mannan sulfates (Kovalenko *et al.*, 1993), a protein from *Mirabilis jalapa* (Kubo *et al.*, 1990) and other plant extracts (reviewed by Verma *et al.*, 1998) were found to be active as resistance inducers. Some of these compounds are also inducers of interferon in animal cells. Apparently, the mechanism that inhibits virus replication in the local lesion area (and induced resistance) can be activated by wide variety of compounds.

D. Salicylic acid

During the HR of *N. tabacum* plants, that possess the *N* gene for resistance to TMV, salicylic acid levels rise markedly (Malamy *et al.*, 1990). It was suggested that salicylic acid plays a role in localization of the virus, as *NN*-genotype transgenic tobacco plants, which have been transformed with a bacterial salicylate hydroxylase gene and, therefore, cannot accumulate salicylic acid, do not limit virus spread. Although the cells of these plants can still undergo HR-type cell death, the plants exhibit a spreading necrosis after TMV inoculation (Darby *et al.*, 2000; Mur *et al.*, 1997), showing that salicylic acid accumulation is required to localize TMV. Cyanide restores the resistance in the transgenic tobacco expressing salicylic acid hydroxylase (Chivasa and Carr, 1998). Interrelated with the role of salicylic acid is the high endogenous level of H_2O_2 during the expression of SAR. Salicylic acid is a known chelator of iron and, thus, it may inhibit several heme-based enzymes including catalase. An increase in the concentration of H_2O_2 may be produced by this inhibition, which has also been reported to result from a binding of SA to a protein with catalase activity. In TMV-infected tobacco leaves, SA levels were reported to increase 10–100-fold and range from 7 to 56 μM (Durner and Klessig, 1995). Expression of a WIPK-activated transcription factor activated upon phosphorylation by the wound-induced protein kinase (WIPK), results in increase of endogenous salicylic acid and pathogen resistance in tobacco plants (Waller *et al.*, 2006). Recently, salicylic acid was assayed *in situ* in tobacco leaves using a genetically modified biosensor strain of *Acinetobacter* sp (Huang *et al.*, 2006).

Treatment of susceptible tobacco with aspirin (acetyl-SA) or SA caused a significant reduction in accumulation of TMV in susceptible tobacco cultivars that do not respond hypersensitively to TMV (Chivasa *et al.*, 1997; White *et al.*, 1983). In leaf mesophyll cells of SA-treated plants replication of TMV is greatly decreased, but not in initially inoculated epidermal cells. However, SA induces resistance to movement between epidermal cells, though SA did not inhibit TMV movement by decreasing the plasmodesmatal size exclusion limit (Murphy and Carr, 2002). SA stimulated formation of callose in *N. glutinosa* infected with TMV, probably affecting the gating capacity of plasmodesmata (Krasavina *et al.*, 2002). Activation of SA-induced protein kinase (SIPK) and WIPK leads to HR-like cell death (Zhang *et al.*, 2000). WIPK is activated in NN-genotype tobacco infected with TMV. PVX-induced gene silencing attenuated *N* gene-mediated resistance (Liu *et al.*, 2003). A WIPK-activating substance was isolated from tobacco leaves and identified as a diterpene. When this compound, natural or chemically synthesized, was applied at nanomolar concentrations to leaves, SIPK was activated and accumulation of transcripts of wound- and pathogen-inducible defense-related genes was enhanced. Treatment of leaves with this diterpene increased resistance to TMV infection (Seo *et al.*, 2003), and in tobacco leaves treated with sulfated fucan oligosaccharides SA accumulated and both local and systemic resistance to TMV was strongly stimulated (Klarzynsky *et al.*, 2003). SA pretreatment primed TMV- infected Xanthi-nc leaves for strong antioxidant induction (Kiraly *et al.*, 2002).

Injecting tobacco (*Nicotiana tabacum*) cv. Xanthi nc plants with nitric oxide (NO)-releasing compounds, caused a significant reduction in the sizes of lesions caused by TMV on the treated leaves and on upper non-treated leaves. The reduction in TMV lesion size was caused by NO released from the NO-releasing compounds. Treatment of tobacco plants with inhibitors of nitric oxide synthase or an NO scavenger attenuated but did not abolish the systemic acquired resistance (SAR) induced by SA (Song and Goodman, 2001).

The structure–activity profile of salicylates and related compounds has been evaluated using an inducible PR protein (PR-1a) and plant resistance to TMV as markers. The seven salicylate derivatives that were the most active TMV resistance inducers were all halogenated in the 3- and/or 5-position. These inducers, listed in order of their strength relative to SA are: 3-chlorosalicylate > 3,5-difluorosalicylate > 3,5-dichloro-6-hydroxysalicylate > 3,5,6-trichlorosalicylate > 5-chlorosalicylate > 5-fluorosalicylate > 3, 5-dichlorosalicylate > 4-fluorosalicylate > 3-fluorosalicylate > 3-chloro-5-fluorosalicylate > 4-chlorosalicylate > salicylic acid (Silverman *et al.*, 2005).

In SA-treated tobacco plants activity of RNA-dependent RNA polymerase (RdRP) increased (Xie *et al.*,2001). Biologically active SA analogs capable of activating plant defense response also induced the RdRP activity, whereas biologically inactive analogs did not. A tobacco gene, NtRDRP1, was isolated and found to be induced both by virus infection and by treatment with SA, suggesting that inducible RdRP plays a role in plant antiviral defense. Similarly, SA induced in *Arabidopsis* an *RdRP* gene with a role in antiviral defense (Yu *et al.*, 2003) [see Carr *et al.* (in press) in volume 76].

E. Some hypotheses that were raised explaining localization

Death of cells in a necrotic lesion may localize or inactivate the virus. This explanation is not satisfactory even in a necrotic local lesion host and even less so in a chlorotic or starch lesion host, as virus particles are found in apparently viable cells outside the necrotic area (Milne, 1966). Furthermore, studies with GFP-tagged TMV (TMV.GFP) have shown that live cells around the necrotic area contain TMV for significant periods of time after lesion formation (Murphy *et al.*, 2001; Wright *et al.*, 2000).

Barrier substances that have been observed to surround local lesions include *inter alia* callose deposition around TMV-induced lesions on Pinto beans (Wu and Dimitman, 1970), calcium pectate in the middle lamella of cells surrounding TMV lesions on *N. glutinosa* (Weintraub and Ragetli, 1964), lignin (Favali *et al.*, 1974) and suberin (Faulkner and Kimmins, 1978) (see also Loebenstein, 1972). However, callose depositions were also observed in infections, which produced systemic necrosis, where the virus does not remain localized, as with tobacco infected with TSWV or *Tobacco rattle virus* (Shimomura and Dijkstra, 1975). On the other hand no callose deposition was observed around TMV-induced starch lesions on cucumber cotyledons, where the infection remains localized without necrosis (Cohen and Loebenstein, 1975). It seems, therefore, that the observed barrier substances are a response to necrotization and not necessarily responsible for the localization.

Inactivation of the virus-coded "transport" or "movement" protein may also help in the localization of the virus. For TMV, this factor was identified as a 30-kDa non-structural protein (Leonard and Zaitlin, 1982). It was shown that the 30-kDa protein accumulated in plasmodesmata, an observation consistent with a role in virus spread (Tommenius *et al.*, 1987). It was also shown that a non-structural protein (P3) of *Alfalfa mosaic virus*, considered to be involved in cell-to-cell spread, is associated with the middle lamella of cell walls (Stussi-Garaud *et al.*, 1987). It was reported that the amount of the 30-kDa protein in the cell wall fraction of TMV-infected Samsun NN tobacco plants decreased sharply as soon as necrosis became visible, compared with that in the systemic host Samsun nn. It was suggested that this might explain why TMV infection becomes localized (Moser *et al.*, 1998). In tobacco NN plants the TMV movement protein alters the gating capacity of plasmodesmata and therefore the efficiency of virus movement (Deom *et al.*, 1991). However, the decrease of transport protein alone does not seem to be responsible for localization, especially as virus particles are found outside the necrotic area (Milne, 1966), and as mentioned above in starch lesions there is no necrotization. Also, the HR in Samsun NN tobacco can be inhibited without affecting the localization of the virus (Takusari and Takahashi, 1979). Furthermore, infection of cowpea by strains of CMV involves a local HR and a localization of infection (inhibition of viral RNA synthesis–IR) (Kim and Palukaitis, 1997). Different combinations of specific sequence alterations in the polymerase gene can separate these responses. Kim and Palukaitis (1997) also showed that IR affects viral RNA synthesis in isolated cells, without HR. The HR on legumes induced by CMV is determined by a region of 243 nt on 2a RNA polymerase gene of CMV, in which two amino acids at positions 631 and 641 in co-determine the induction of HR on legumes (XiaoRong *et al.*, 2003).

CMV induces local lesions in inoculated leaves of *Chenopodium amaranticolor* but fail to doso if lacking either the 3a movement protein or the coat protein. Cytological analysis showed that both viral-encoded proteins are required for cell-to-cell movement of the virus and the simultaneous appearance of cellular necrosis. In the absence of either or both proteins, infection was confined to single, non-necrotized, epidermal cells. Apparently, viral movement out of the initially infected epidermal cell is required for the induction of cell death (Canto and Palukaitis, 1999).

F. The *N* gene

In *N. glutinosa* and *N. tabacum cultivars* that contain the *N* gene, TMV does not spread but remains localized in a lesion of several hundred cells. The *N* gene was originally transferred from *N. glutinosa* to *N. tabacum* via an interspecific hybrid, *N. digluta* (Clausen and Goodspeed, 1925; Holmes, 1954). Holmes (1938) showed that resistance to TMV was controlled by a single dominant gene–which he termed *N*. However, some other genes were also introgressed into tobacco from *N. glutinosa* together with the *N* gene locus (Holmes, 1938). The genetic and breeding history was described by Dunigan *et al.* (1987). This gene was of major importance in infectivity assays, as the number of necrotic lesions was in a certain proportion to the virus content (Kleczkowsky, 1950). In the early 1990s, the *N* gene was the first plant virus resistance gene to be isolated by transposon tagging, using the maize activator transposon (Dinesh-Kumar *et al.*, 1995; Whitham *et al.*, 1994). The *N* gene encodes a protein of 131.4 kDa, which has three domains: an N-terminal TIR domain similar to that of the cytoplasmic domain of the *Drosophila* Toll protein and the interleukin-1 receptor in mammals; an NBS site, and four imperfect leucine-rich repeats (LRR). The *N* gene thus belongs to the TIR-NBS-LRR class of *R* genes. The *N* gene encodes two transcripts, N_S and N_L, generated via alternative splicing of the alternative exon (AE) present in the intron III. The N_S transcript, predicted to encode the full-length N protein containing the Toll-IL-1 homology region, NBS, and LRR is more prevalent before and for 3 h after TMV infection. The N_L transcript, predicted to encode a truncated N protein (N^{tr}) lacking 13 of the 14 repeats of the LRR, with a deduced molecular weight of 75.3 kDa, is more prevalent 4–8 h after TMV infection. The ratio of N_s to N_t before and after TMV inoculation is critical to achieve complete resistance. The N^{tr} protein is identical to the amino terminal portion of the N protein, with an additional 36 amino acids at the C-terminus. Plants harboring a cDNA-N_S transgene, capable of encoding an N protein but not an N^{tr} protein, fail to exhibit complete resistance to TMV. Transgenic plants containing a cDNA-N_S-bearing intron III (without introns I, II, and IV) and containing 3′ *N*-genomic sequences, encoding both N_S and N_L

transcripts, exhibit complete resistance to TMV. These results suggest that both *N* transcripts and presumably their encoded protein products are necessary to confer complete resistance to TMV. However, deletion of the 70 bp AE with flanking splice acceptor–donor sites, resulted in delayed HR upon infection with TMV and the virus continued to spread, resulting in systemic HR (Dinesh-Kumar and Baker, 2000). In addition to the *N* gene, other TIR-NBS-LRR *R* genes, such as L6, RPP5 and RPS4 also encode two or more transcripts. The biological role of these genes is presently unknown (Marathe *et al.*, 2002).

The *N* gene shows some structural similarities to a number of other plant genes for disease resistance, as the *RPS*2 and *RPM*1 genes for resistance to *Pseudomonas syringae* pathovars in *Arabidopsis* (Bent *et al.*, 1994) and *Prf* for resistance to *P. syringae* in tomato (Salmeron *et al.*, 1994). There seems therefore to be common structural elements in genes for resistance to different types of pathogens as the NBS and LRR motifs.

The presence of TIR, NBS, and LRR domains are all necessary for proper *N* function (Dinesh-Kumar *et al.*, 2000). Most in frame deletion mutants in the *N* gene abolished resistance to TMV. Also, some amino acid substitutions within the TIR domain caused a complete loss of the *N* function, and the plants developed systemic HR. It was concluded that mutations that affect *Drosophila* Toll or human IL-IR signaling also affect the *N*-mediated response to TMV (Dinesh-Kumar *et al.*, 2000). The NBS domain of the N protein has amino acid homology with regions of cell-death proteins (Arvind *et al.*, 1999; Van der Biezen and Jones, 1998). The NBS domain is also found in elongation factors and members of G-protein families (Saraste *et al.*, 1990). These serve as molecular switches in growth and differentiation. Point mutation in some subdomains of the NBS led to loss of resistance. In plants expressing some of the mutant N proteins timing of HR appearance and size of lesions was normal, but TMV spread systemically through the plant, causing death of the plant within 5–7 days (Dinesh-Kumar *et al.*, 2000). It was shown that introgression of the *N* gene coincided with introgression of DNA-bearing restriction enzyme fragments of *N. glutinosa* origin (Whitham *et al.*, 1994). The *N* protein (to the best of our knowledge) has so far not been purified and its way of function has not been determined. It is possible that the *N* protein may trigger an intracellular signal transduction cascade and induces a variety of defense and signaling proteins, as for example the IVR protein (see below). It may be that the *N* protein activates the HR but activation of resistance requires another gene, similar to the *HRT* and *RRT* genes in TCV resistant *Arabidopsis* (Kachroo *et al.*, 2000) (see below).

The pathway by which signals from the *N* gene product are transmitted is largely unknown. Yoda *et al.* (2002) identified seven tobacco genes that are associated with HR upon TMV infection.

Transcriptional induction of one of these, which encodes a novel WRKY transcription factor, is independent of SA. Its full-length cDNA of 1346 bp encoded a polypeptide consisting of 258 amino acids. The deduced protein contained a single WRKY domain, a Cys_2His_2 zinc-finger motif and a leucine-zipper motif, showing high similarity to WIZZ, a member of the family of WRKY transcription factors in tobacco. This indicated the presence of salicylic acid-independent pathways for HR signal transduction, in which a novel type of WRKY protein(s) may play a critical role for the activation of defense (Yoda *et al.*, 2002).

The *N* gene has been transferred to tomato, where it confers resistance to TMV (Whitham *et al.*, 1996). While characterizing the N-TMV signaling pathway and its components in tomato a mutant, *sun1-1* that is defective in the N-TMV signal was identified. The sun1-1 mutation affects a tomato homolog of Arabidopsis EDS1. Tomato EDS1 lies upstream of salicylic acid and is required for resistance mediated by TIR class resistance genes and the receptor-like resistance gene *Ve*. Relative to wild-type susceptible plants, *sun1-1* plants show enhanced susceptibility to TMV, providing evidence that the intersection of general and resistance gene-mediated pathways is conserved in Solanaceous species. In tomato, EDS1 is important for mediating resistance to a broad range of pathogens (viral, bacterial, and fungal pathogens), yet shows specificity in the class of *R* genes that it affects (TIR-NBS-LRR) (Hu *et al.*, 2005).

The N protein, either by itself or in a protein complex, is hypothesized to specifically recognize the 50-kDa C-terminal helicase domain of the TMV replicase protein (Abbink *et al.*, 1998; Padgett *et al.*, 1997) and trigger a signal transduction cascade leading to induction of HR and restriction of virus spread.

In transgenic *Nicotiana benthamiana*, containing the tobacco *N* gene, it was shown that the *Rar1-*, *EDS1-*, and *NPR1/NIM1*-like genes are required for the N-mediated resistance (Liu *et al.*, 2002a). The *Rar1* gene encodes a protein with a zinc finger motif, which is required for the function of *N*. It was shown that *N. benthamiana Rar*1 (NbRar1) protein interacts with NbSGT1 a highly conserved component of an E3 ubiquitin ligase complex involved in protein degradation. It also interacts with the COP9 signalosome, a multiprotein complex involved in protein degradation via the ubiquitine-proteasome pathway. Suppression of NbSGT1 and NbSKP1 (an SCF protein complex component) in *NN* transgenic plants by virus-induced gene silencing (VIGS) resulted in the loss of resistance to TMV. This indicates that NbSGT1 and NbSKP1 are required for *N* function (Liu *et al.*, 2002b). It is interesting to note that genetic studies in barley suggested that *Rar1* functions downstream of pathogen perception and upstream of H_2O_2 accumulation and host cell death (Shirasu *et al.*, 1999).

Infection of resistant tobacco plants that carry the *N* resistance gene with TMV leads to the activation of two tobacco mitogen-activated protein kinases (MAPK), SIPK, and WIPK (Zhang and Klessig, 1998). WIPK gene transcription is regulated by phosphorylation and dephosphorylation events and may accelerate the HR cell death (Liu *et al.*, 2003).

A gene with a function similar to that of a resistance gene named NH has been cloned from *N. tabacum* cv. Xanthi nn plants. The coding region of NH is 5.028 base pairs (bp) long and has 82.6% nucleotide identity with the *N* gene. In contrast to the *N* gene, the *NH* gene lacks intron 4 and does not have sites for alternative splicing of intron 3. Analysis of its sequence revealed that NH belongs to the TIR/NSB/LRR gene class. It was suggested that this gene, homologous to the *N* gene, plays a role in the HR response of some *Nicotiana* species (Stange *et al.*, 2004).

G. Other host-virus HR responses

HR in potato to PVX is controlled by genes *Nb* (Rb) and *Nx* (*Rx*), which have been mapped to a gene cluster in the upper arm of chromosome V (De Jong *et al.*, 1997), and to a region of chromosome IX (Tommiska *et al.*, 1998), respectively. The same region of chromosome IX contains the gene Sw-5 for resistance to tomato spotted wilt tospovirus in tomato. The *Nx*-mediated resistance is elicited by the PVX coat protein gene (Kavanagh *et al.*, 1992). Most PVX strains induce HR on potato carrying the *Rx* gene, which depends on the presence of a threonine residue at position 121 of the PVX coat protein. Elicitation of lesions on *Gomphrena globosa* also required presence of the threonine residue at position 121 of the coat protein (Goulden and Baulcombe, 1993). However, the *Rx* gene confers resistance to PVX in potato by arresting virus accumulation in the initially infected cell without an HR response (Köhm *et al.*, 1993). Rx shows similarities with NBS-LRR class of *R* genes (Bendahmane *et al.*, 1999) (see Chapter 1). Also, many PVY strains induce necrosis (HR) in potato while spreading systemically in the plant (Jones, 1990). Again it is evident that HR and inhibition of virus replication are two separate processes.

The coding region of a potato gene termed Y-1 was found to be structurally similar to the tobacco *N* gene. It is located at the distal end of chromosome XI in potato *Solanum tuberosum* subsp. *andigena*. This gene also belongs to the TIR-NBS-LRR class and has 57% identity at the amino acid level as predicted by the sequence with that of the *N* gene. The *R* gene rich region of chromosome XI is syntenic (*two or more genes being located on the same chromosome whether or not there is demonstrable linkage between them*) in potato and tobacco (Vidal *et al.*, 2002). It contains the *N* gene and genes homologous to *N* with unknown functions in potato (Hamäläinen *et al.*, 1998). The coding region of the gene is 6187-bp long.

Leaves of transgenic potato plants expressing Y-1 developed necrotic lesions upon infection with PVY, but no resistance was observed, and plants became systemically infected by PVY (Vidal *et al.*, 2002).

The dominant gene *HRT* in TCV-resistant *Arabidopsis* encodes a member of the class of R proteins that contain a leucine zipper, an NBS, and leucine-rich repeats and therefore belongs to the LZ-NBS-LRR class of genes. Inoculation of TCV onto resistant *Arabidopsis* leads to a HR. Other ecotypes of *Arabidopsis* do not give an HR but allow systemic spread of the virus (Li and Simon, 1990). *HRT* was cloned and conferred HR TCV in transgenic plants (Cooley *et al.*, 2000). *HRT* shares extensive sequence similarity with members of the *RPP8* gene family, which confer resistance to the oomycete pathogen *Peronospera parasitica*. Transgenic plants expressing *HRT* developed an HR but generally remained susceptible to TCV because of a second gene, *RRT*, which regulates resistance to TCV (Kachroo *et al.*, 2000). However some of the transgenic lines that were resistant did not develop a normal HR.

Some soybean lines resistant to *Soybean mosaic virus* (SMV) due to the resistance gene *Rsv1* show an HR response to infection (Hill, 2003). This gene is within a cluster with other resistance (R) genes against *Pseudomonas syringae*, *Phytophthora sojae*, a root-knot nematode and against *Peanut mottle* and *Peanut stripe virus* (mentioned in Penuela *et al.*, 2002). Genes in this cluster code for proteins belonging to the NBS LRR superfamily, with a coiled-coil motif (non-TIR) (Penuela *et al.*, 2002). In addition Wang *et al.* (2003) screened a soybean cDNA library and isolated resistance gene analogs with an NBS domain. A cDNA of 2533 bp in length was obtained that coded for a polypeptide of 636 amino acids with TIR and NBS domain, which was similar to the *N* gene of tobacco, with sequence identity of 28.1%. The expression of this gene could be induced by exogenous salicylic acid.

In sweet pepper, the *Tsw* gene, originally described in *Capsicum chinense*, has been widely used as an efficient gene for inducing an HR-derived TSWV resistance (Lovato *et al.*, 2008).

H. The *N′* gene

The *N′* gene, originating from *N. sylvestris*, controls the HR induced by many strains of TMV, except U1 and OM, which spread systemically and produce mosaic symptoms in *N′* containing plants. Mutants that induce necrosis can easily be isolated from infections causing systemic mosaic symptoms (Culver *et al.*, 1991). This system is a nice one to demonstrate cross protection. Alterations in the structure of the TMV coat protein affect the elicitation of the *N′* gene HR in *N. sylvestris*. Two specific amino acid substitutions within the virus coat protein were responsible for host recognition and HR elicitation (Culver and Dawson, 1989). Five amino acid substitutions were identified to elicit HR (Culver *et al.*, 1991). The three dimensional structure of coat protein is critical to induce the HR response, either directly through specific structural motifs or indirectly via alterations in coat protein assembly (Culver, 2002). Substitutions eliciting HR were within, or would predictably interfere with, interface regions between adjacent subunits in ordered aggregation of the coat protein. Substitutions that did not elicit the HR were either conservative or located outside the interface region. Radical substitutions that predictably disrupted coat protein tertiary structure prevented HR elicitation (Culver *et al.*, 1994). Transgenic plants expressing elicitor coat proteins developed necrotic patches that eventually coalesced and collapsed entire leaves, demonstrating that expression of elicitor coat proteins independently of viral replication can induce the HR in *N. sylvestris* (Culver and Dawson, 1991). A point mutations in the mutant at nucleotides 6157 (cytosine to uracil) produced local lesions on *N. sylvestris*, and mediating the outcome of infection in *N. sylvestris* (Knorr and Dawson, 1988).

I. Inhibitor of virus replication (IVR)

Localization of TMV in tobacco, containing the *N* gene, is associated with the presence of a protein with antiviral properties named 'inhibitor of virus replication (IVR) (Gera and Loebenstein, 1983; Gera *et al.*, 1990; Loebenstein and Gera, 1981). IVR was released into the medium of TMV-infected protoplasts derived from (Samsun NN) tobacco. When added to the medium up to18 h after inoculation, IVR inhibited virus replication in protoplasts derived from both resistant Samsun NN exhibiting local lesions and systemically infected susceptible *N. tabacum* cv. Samsun plants (Samsun nn).

IVR inhibited TMV in protoplasts and leaf disks, the effect being dose responsive, reaching 70–80% with 10 units of IVR (Gera *et al.*, 1986). IVR also inhibited PVX, PVY, and CMV in leaf disks from different hosts, indicating that IVR is neither host nor virus specific. IVR inhibited TMV replication in intact leaves when applied by spraying to tobacco and tomato plants and CMV in cucumbers. IVR was found to be sensitive to trypsin and chymotrypsin, but not to RNase. And its ability was abolished by incubation at 60°C for 10 min (Gera and Loebenstein, 1983).

IVR was also obtained from the intracellular fluid of hypersensitive tobacco leaves infected with TMV (Spiegel *et al.*, 1989). Production of IVR by infected protoplasts and by intact Samsun NN plants was suppressed almost completely when exposed to 35°C, leading to accumulation of TMV (Gera *et al.*, 1993). Also treatment of Samsun NN protoplasts with actinomycin D and chloramphenicol 5 or 24 h after their inoculation decreased IVR in the incubation medium to close to zero, concomitant with a marked increase of TMV in the protoplasts (Gera *et al.*, 1983). No increase was observed when TMV-infected protoplasts of Samsun (susceptible to systemic infection by TMV) were incubated in the presence of these antimetabolites. A ca. 23-kDa protein band was always associated with samples of crude protoplast IVR, tissue-IVR and IVR purified from induced-resistant tissue; this protein was absent in samples of uninfected plant tissue and protoplasts derived from them. Purification of the 23-kDa protein from SDS-polyacrylamide gels yielded a molecule with antiviral properties in biological tests (Gera *et al.*, 1990). Antibodies against the IVR protein neutralized its antiviral activity and enabled immunodetection of the 23-kDa protein (Gera and Loebenstein, 1988; Gera *et al.*, 1990).

Sequence analysis of clone NC330 indicated that the C-terminus of the deduced protein is highly acidic, rich in aspartic acid and glutamic acid, hydrophobic and with a helical structure (Akad *et al.*, 1999). NC330 Protein (accession CAA08776) motif analysis in silico showed the presence of six sites typical for protein kinase one site for N-glycosylation, two N-merystylation sites and LRR, but it is mainly a tetratricopeptide repeat (TPR) protein. These motifs are known to be involved in protein–protein interactions. It is worthwhile to note that TPR motifs are present in many proteins including inducible interferons (Der *et al.*, 1998; Zhang and Gui, 2004).

A direct involvement of TPR motif in plant resistance to pathogens was recently described with RAR1 interactor protein. RAR1 is an early convergence point in a signaling pathway engaged by multiple *R* genes (Azevedo *et al.*, 2002). It was proposed that two TPR proteins RAR1 and SGT1 function with HSP90 in chaperoning roles that are essential for disease resistance (Hubert *et al.*, 2003; Takahashi *et al.*, 2003).

When NC330 was compared with other proteins in the GenBank marked identities were observed with two putative proteins from *Arabidopsis* (accession NP_850309 and A84813). Both proteins had 78% identity with the NC330 protein, indicating that NC330 is a well-preserved protein. The NC330 transcript homolog was found in several Expressed Sequence Tags (EST) from plants, mainly plants at different development stages or stressed plant tissue. This indicates that NC330 is essential for development (as *SPY* and *OGT* genes) and in resistance responses (as *RAR1*, *SGT1*, and *HRT* genes).

Transformation of *N. tabacum* cv. Samsun nn, in which TMV spreads systemically, with NC330, encoding an IVR-like protein, resulted in a number of transgenic plants, expressing variable resistance to TMV and the fungal pathogen *Botrytis cinerea* (Akad *et al.*, 2005). Also tomato plants transformed with NC330 resulted in a number of transgenic plants expressing resistance to *B. cinerea* (in preparation).

J. IVR-like compounds associated with other resistance responses

The interspecific hybrid of *N. glutinosa X N. debneyi* is highly resistant to TMV (Ahl and Gianinazzi, 1982). Following inoculation, only a small number of tiny lesions develop from which only an extremely low level of infectivity can be recovered. A specific band corresponding to a 23-kDa protein was consistently observed in PAGE of crude hybrid IVR, both from TMV-inoculated and uninoculated hybrid plants. This band reacted specifically in immonoblots with IVR antiserum and crude extracts obtained from inoculated or uninoculated leaves of the hybrid gave positive reactions with IVR antiserum in agar-gel diffusion tests. The precipitation lines fused without spur formation with the precipitation lines obtained between protoplast IVR and the antiserum. Based on these criteria IVR from the hybrid was indistinguishable from IVR obtained from Samsun NN (Loebenstein *et al.*, 1990).

An IVR-like protein (about 23 kDa) is constitutively produced in a resistant pepper cultivar, but not in a susceptible one. This constitutive production of the IVR-like protein in this host may be responsible for its high resistance to TMV (Gera *et al.*, 1994).

A substance(s) which inhibits virus replication (Inhibitor from Green Islands IGI), was released by protoplasts obtained from green island leaf tissue of tobacco, cv. Xanthi-nc, infected with CMV. It was also obtained directly from green island tissue. IGI inhibited both CMV and TMV replication in protoplasts and leaf tissue disks, with the degree of inhibition being dependent upon concentrations applied. The IGI was partially purified to yield two active fractions with molecular weights of about 26 kDa and 57 kDa. IGI appeared to be quite similar to IVR since both posses' similar serological determinants. Fractions with similar activity were also obtained from green island leaf tissue of tobacco cv. Samsun, infected with CMV. However, these fractions differed serologically from both the IVR and IGI obtained from Xanthi-nc (Gera and Loebenstein, 1988).

II. INDUCED RESISTANCE

The term *induced resistance* will be used for resistance developing in non-invaded tissue after other parts of the plant have been infected with a virus and activation of resistance by non-viral agents. Many reviews have been written on various aspects of induced resistance and from different viewpoints; one of the more recent ones was by Gilliland *et al.* (2006).

A. Induced by viruses

The first observation that infection in one part of the plant induces resistance in other non-invaded tissues was reported by Gilpatrick and Weintraub (1952) in *Dianthus barbatus* inoculated with *Carnation mosaic virus*. More detailed studies were done by Yarwood (1960) with TMV in beans and by Ross (1961a) with TMV in tobacco. The zone around the primary lesions became resistant to a second inoculation with TMV. This was termed *local induced (acquired) resistance* (LAR). Subsequent studies showed that not only the leaf tissue adjacent to the lesion became resistant but other tissues, as well—*systemic acquired resistance* (SAR) (Ross, 1961b). This distinction seems to be one of convenience, as it depends only on the distance of the resistant tissue from the primary infection. In the resistant zones, both LAR and SAR, after challenge inoculation with a virus that causes local lesions, both the number and size of lesions is reduced significantly. The resistance is not specific, and for example resistance induced by TMV in NN-genotype tobacco is not specific for TMV, but also resistant when challenged with TNV, TRSV and TomRSV. In tobacco *N* gene-mediated resistance is highest when tests are carried out at 20–24°C; and no resistance becomes evident at temperatures close to 30°C (reviewed by Loebenstein, 1972). Further classical examples of SAR in other plants include cucumber (*Cucumis sativus*), common bean (*Phaseolus vulgaris*), rice (*Oryza sativa*), and *Arabidopsis*.

SAR is conserved across diverse plant families and is effective against a broad range of viral, bacterial, and fungal pathogens (reviewed in Sticher *et al.*, 1997). SAR has also been reported in lesion—mimic mutants of *Arabidopsis* and barley (Dietrich *et al.*, 1994; Wolter *et al.*, 1993). In tomato cv. Jiafen 16, localized TMV infection could induce cell death in the uninoculated parts of the plants, where enzyme-linked immunosorbent assay (ELISA) showed no spreading virus (Zhou *et al.*, 2008). Tobacco plants expressing vacuolar and apoplastic yeast-derived invertase, develop spontaneous necrotic lesions similar to the hypersensitive responses caused by avirulent pathogens. This was accompanied by increased resistance to PVY, as measured by decreased vira1 spread and reduced multiplication in systemic leaves of the transgenic plants (Herbers *et al.*, 1996).

Activation of six groups of genes that are expressed relatively early in the inoculated leaves of tobacco (*N. tabacum*) resisting infection by TMV, is independent of salicylic acid and ethylene. Induction of all these genes was subsequently detected in the uninoculated leaves; thus, their expression is associated with the development of both LAR and SAR. Exogenously applied SA induced these genes transiently (Guo *et al.*, 2000). Several peptides originating from pathogens can activate the plant innate immune response, including fungal and bacterial elicitors. Thus, a

23-aa peptide isolated from extracts of Arabidopsis leaves and called AtPep1, exhibited characteristics of an endogenous elicitor of the immune response (Huffaker *et al.*, 2006).

B. SAR and salicylic acid

Ward *et al.* (1991) showed that the onset of SAR correlates with the coordinate induction of nine classes of mRNAs. Salicylic acid, a candidate for the endogenous signal that activates the resistant state, induces expression of the same "*SAR* genes." These genes encode PR-proteins, chitinase and glucanase. This, however, is probably coincidental, as so far no evidence has been provided that they are active in virus-induced SAR or localization of viruses. They are mainly induced in virus infection that do cause necrosis and may therefore be a host response to necrosis. Accumulation of SA is required for SAR, but only in the signal-perceiving systemic tissue: Grafting experiments showed that tobacco leaves infected with TMV could transmit a SAR signal despite the presence of bacterial salicylate hydroxylase (Vlot *et al.*, 2008); and expression of this SA-degrading enzyme in systemic tissue abolished SAR signal perception (Vernooij *et al.*, 1994). Park *et al.* (2007) showed that the SA-derivative methyl salicylate (MeSA) is not degraded by SH, accumulates in *NahG* transgenic tobacco, and acts as a long-distance mobile signal for SAR. In addition to serving as an endogenous SAR signal, MeSA can serve as an airborne signal that is emitted from infected plants and induces defense gene expression in neighboring wild-type plants (Shulaev *et al.*, 1997). Ward *et al.* (1991) reported that a synthetic immunization compound, methyl-2,6-dichloroisonicotinic acid, also induces both resistance and *SAR* gene expression. Yang and Klessig (1966) suggested that one mechanism of SA action is to inhibit catalase and ascorbate peroxidase, thereby elevating endogenous H_2O_2 levels and generating salicylate free radicals. The increased levels of H_2O_2 and/or products of reactions induced by salicylate free radicals such as lipid peroxide may act as signals for the activation of plant defense genes. Salicylic acid (SA) has been associated with the activity of NPR1, a protein that interacts with transcription factors to regulate the expression of defense-related *PR* genes (Zhang *et al.*, 1999). The expression of the *PR-1* gene has been used, especially in tobacco and *Arabidopsis*, as a marker for these SA-mediated responses. SAR apparently does not require long-distance translocation of SA (Truman *et al.*, 2007) and jasmonates possibly act as the initiating signal for classic SAR. Thus, jasmonic acid (JA), but not SA, was found to accumulate in phloem exudates of leaves challenged with an avirulent strain of *Pseudomonas syringae*. Recently, however, Attaran *et al.* (2009) published that jasmonates signaling is not essential for SAR in *Arabidopsis*. Another set of responses is dependent on the production of either JA or ethylene and involves the gene *PDF1.2*. These latter responses are effective against fungal and oomycete pathogens (Clarke *et al.*, 2000). A lipid transfer protein seems also to be involved in systemic resistance signaling in *Arabidopsis*. Thus, a mutation affecting the lipid-transfer protein DIR1 (Defective in induced resistance 1) renders *Arabidopsis* incapable of generating/transmitting a functional SAR signal, but does not affect resistance in the inoculated leaf (Maldonado *et al.*, 2002).

Evidence is accumulating that different defense pathways interact such that induction of one pathway may inactivate the other (Traw *et al.*, 2003) resulting in resistance to different organisms.

The expression of the *PR-1* gene has been used, especially in tobacco and *Arabidopsis*, as a marker for these SA-mediated responses. Chitosan treatments induced resistance to challenge by *Tomato bushy stunt virus* (Faoro *et al.*, 2001) and reduced both size and number of lesions caused by TNV in *Phaseolus vulgaris* (forming microlesions). Apparently the chitosan-induced microlesions are responsible for the observed high local resistance, meanwhile generating signals for the induction of SAR (Faoro and Iriti, 2006). This resistance was associated with a network of callose deposits, micro-oxidative bursts and micro-hypersensitive responses (micro-HRs). Electrophoresis of genomic DNA extracted from cultured cell after 48 h treatment showed internucleosomal fragmentation, visualized as a distinct ladder of DNA bands corresponding to oligonucleosomal units (Iriti *et al.*, 2006). SAR induced by chito-oligosaccharides strongly suppressed the expression of TMV coat protein gene in tobacco (Shang *et al.*, 2007). In an EST analysis of resistance to SMV in soybean it was observed that the genes associated with SAR were the most diverse and abundant (Liu-Chun *et al.*, 2005).

Probenazole induces SAR in tobacco through salicylic acid accumulation (Nakashita *et al.*, 2002); and sulfated beta-1,3 glucan PS3 activates the salicylic acid (SA) signaling pathway in infiltrated tobacco and *Arabidopsis* leaf tissues, but does not induce SAR to TMV (Menard *et al.*, 2005). This may indicate that accumulation of SA does not always lead to SAR.

An SA-binding protein 2 (SABP2), is present in low abundance in tobacco and specifically binds SA with high affinity. Sequence analysis predicted that SABP2 is a lipase belonging to the α/β fold hydrolase super family. Silencing of SABP2 expression suppressed local resistance to TMV and development of SAR (Kumar and Klessig, 2003). Expression of a syn, but not a nat, SABP2 gene restored SAR (Kumar *et al.*, 2006).

Leaves of infected *Arabidopsis* mutants fail to emit a conserved SAR signal that induces defense gene expression or pathogen resistance in *Arabidopsis*, tomato, and/or wheat. However, petiole exudates from infected plants restore systemic defense signaling of comparable exudates from the mutants indicating that a glycerolipid-derived factor may interact with DIR1 to trigger SAR.

Another potential lipid-derived SAR signal is the oxylipin-derived defense hormone jasmonic acid which might be an early signal establishing systemic immunity. Jasmonic acid or ethylene are effective against fungal and oomycete pathogens (Clarke *et al.*, 2000). Evidence is accumulating that different defense pathways interact such that induction of one pathway may inactivate the other (Traw *et al.*, 2003) resulting in resistance to different organisms.

Inducible resistance pathways are costly to the plant, and their constitutive expression in the absence of a pathogen may result in a decrease in fitness (Heidel *et al.*, 2004). Thus, several mutations (*npr1*, *cpr1*, *cpr5*, and *cpr6*) and two transgenic genotypes (*NPR1-L* and *NPR1-H*) affecting different points of the SAR signaling pathway associated with pathogen defense, affected the fitness of *Arabidopsis*. Constitutive activation of SAR by *cpr1*, *cpr5*, and *cpr6* generally decreased fitness in the field.

Non-lethal concentrations of antimycin A (AA) or cyanide (CN–) were found to induce resistance to TMV in susceptible tobacco (Chivasa *et al.*, 1997) and to the related tobamovirus *Turnip vein-clearing virus* (TVCV) in *Arabidopsis* (Wong *et al.*, 2002), without inducing *PR1* gene expression. Based on this pharmacological evidence, a model was proposed in which the signal transduction pathways involved in virus resistance separate downstream of SA; one branch (sensitive to salicylhydroxamic acid) leads to resistance to viruses, the other (insensitive to salicylhydroxamic acid) to the induction of PR proteins and to bacterial and fungal resistance (Gilliland *et al.*, 2006).

From a study with ethylene-insensitive (Tetr) tobacco, it was concluded that ethylene perception is required to generate the systemic signal molecules in TMV-infected leaves that trigger SA accumulation, defence gene expression, and SAR development in uninfected leaves (Verberne *et al.*, 2003).

In potato cv. Desiree inoculated with PVX salicylic acid biosynthesis and expression of several defence genes including PR-1 and glutathione-*S*-transferase, which are involved in ethylene and reactive oxygen species dependent signaling, were highly up-regulated in upper-uninoculated (systemic) leaves, compared with mock-inoculated controls. Moreover, the beta-phenyl ethylamine-alkaloids tyramine, octopamine, dopamine, and norepinephrine were highly induced upon infection. β-phenylethylamine-alkaloids can contribute to active plant defence responses by forming hydroxycinnamic acid amides (HCAA), which are thought to increase cell wall stability by extracellular peroxidative polymerization. Expression of tyramine-hydroxycinnamoyl transferase (THT) and apoplastic peroxidase (POD) was highly induced upon PVX infection in systemic leaves, which suggests synthesis and extracellular polymerization of HCAA (Niehl *et al.*, 2006).

Nie (2006) investigated the effects of SA and ACC on the systemic development of symptoms induced by a severe isolate of PVY group N:O (PVYN:O) in tobacco. Treatment of seedlings with SA delayed the virus-induced necrosis in stems by 1–2 days. SA, not ACC, also significantly suppressed the symptom severity in stems. However, neither SA nor ACC treatment affected the partial recovery phenotype exhibited in the latterly emerged upper parts of the plants. Further analysis indicated that the accumulation of PVY was retarded by SA at the early stage of infection, and the effects were more profound in stems than leaves. Nie (2006) suggested that SAR plays a key role in suppressing PVYN: O-induced symptom development through SA-mediated and ethylene-independent pathways. The symptom suppression was correlated with reduced replication/accumulation of virus at the early stage of infection. The results also suggest that neither SA nor ethylene plays a role in the recovery phenotype.

Racemic β-aminobutyric acid (BABA) is a non-protein amino acid that has a broad spectrum of activity against diseases caused by necrotrophic fungi, bacterial, and viral pathogens in several crops. When applied to tobacco leaves at 10 mM, the compound caused local HR-like lesions, as a consequence of the oxidative burst and related events (cell death, lipid peroxidation, and callose formation), followed by a local and systemic increase in SA content and expression of PR-1. The enhancement of resistance to TMV appeared to depend on the SA-mediated pathway as it was abolished in transgenic NahG plants (Siegrist *et al.*, 2000).

Induced resistance studies have focused on RNA viruses, but were also reported with a DNA virus CaMV. Both SA and CN treatment of *Arabidopsis* delayed the appearance of symptoms caused by CaMV. SA treatment severely depressed the accumulation of CaMV DNA, and probably as a consequence the 35S and 19S RNA species. The SA-induced resistance against CaMV occurred in directly inoculated tissue indicating that replication, or possibly cell-to-cell movement, is affected (Gilliland *et al.*, 2006). SA can inhibit cell-to-cell movement of TMV (GFP) in the epidermis, and interferes with TMV (GFP) replication in the mesophyll cell (Murphy and Carr, 2002).

It seems that SAR and gene silencing are connected. Thus, the 2b protein of CMV, which is a suppressor of gene silencing, also interferes with SA-induced resistance to viruses (Ji and Ding, 2001). Other evidence for this includes the fact that RNA-dependent RNA polymerases (RdRPs) are also associated with antiviral defense. In *N. benthamiana* one RdRP was similar in sequence to SDE1/SGS2 required for maintenance of transgene silencing, whereas the second, named NbRdRP1m, was >90% identical in sequence to the SA-inducible RdRP from *N. tabacum* required for defense against viruses. NbRdRP1m expression was induced by SA treatment or challenge with TMV, but the gene and transcript sequences differed from those of other SA-inducible RdRPs in that they contained a 72-nt insert with tandem in-frame stop codons in the 5′ portion of the ORF. *N. benthamiana* plants transformed with an SA-inducible RdRP gene from *Medicago truncatula* were more resistant to infection by TMV, TVCV, and Sunnhemp mosaic virus (members of Tobamovirus genus), but not to CMV and PVX (Yang *et al.*, 2004).

C. Synthetic and natural compounds that induce resistance

The idea of protecting crops by chemical activation of their own defenses appears interesting but so far is not being used in practice. The progress achieved with compounds emulating the role of endogenous inducers indicates that these treatments may lead to results, which then might be integrated into pest management practices.

Many compounds including those mentioned in Section D induce SAR. In addition the mammalian 2′5′ oligoadenylate system (2-5A system) expressed in tobacco plants exhibited resistance to CMV (Honda *et al.*, 2003).

A strain of the non-pathogenic rhizobacterium EXTN-1 from *Bacillus amyloliquefaciens* is capable of eliciting broad-spectrum induced systemic resistance (ISR) in tobacco and *Arabidopsis* that is phenotypically similar to pathogen-induced SAR. (The term ISR was used to prime systemic resistance against bacterial and fungal pathogens by a mechanism that is not SA dependent.) Strains of plant growth-promoting rhizobacteria (PGPR) (as *Bacillus pumilus* SE34 and *Pseudomonas fluorescens* 89B-61) were also effective in reducing CMV and TSWV damage in tomato (Zehnder *et al.*, 2001). SAR to TMV in tobacco was elicited by 90 kDa extracellular protein purified from the culture filtrate of *Phytophthora boehmeriae* (Zhang *et al.*, 2002); and the *Phytophthora megasperma* glycoprotein elicitin induces LAR and SAR to TMV in tobacco (Cordelier *et al.*, 2003). Also, the elicitor protein PemG1 from *Magnaporthe grisea*, when expressed in tobacco plants improved the resistance to TMV (Mao *et al.*, 2008).

Specific calmodulin (CaM) isoforms were activated by infection or pathogen-derived elicitors, and participated in Ca^{2+}-mediated induction of plant disease resistance. Soya bean CaM (SCaM)-4 and SCaM-5 genes, which encode for divergent CaM isoforms, were induced in cell cultures within 30 min by fungal elicitors or pathogens. The pathogen-triggered induction of these genes specifically depended on the increase of intracellular Ca^{2+} levels. Expression of SCaM-4 and SCaM-5 in transgenic tobacco plants triggered spontaneous induction of lesions and induced an array of SAR-associated genes. The transgenic plants exhibited enhanced resistance to TMV. CaM isoforms are components of a SA-independent signal transduction chain leading to disease resistance (Won *et al.*, 1999).

"Bion" (trade name for ASM also called benzo (1,2,3) thiadiazole-7-carboxylic acid (BTH) (Oostendorp *et al.*, 2001), is a synthetic functional analogue of salicylic acid developed for use in a novel strategy for crop protection through induction of SAR. In monocots, activated resistance by BION is very long lasting, while the lasting effect is less pronounced in dicots. ASM protected cantaloupe against a fungal pathogen and CMV. ASM induced the systemic accumulation of chitinase, a marker protein for SAR, in both greenhouse and field grown seedlings (Smith-Becker *et al.*, 2003). ASM also activated SAR in flue-cured tobacco to TSWV. ASM restricted virus replication and movement, and as a result reduced systemic infection. Tobacco plants treated with increased quantities of ASM showed increased levels of SAR (Mandal *et al.*, 2008). ASM pretreatment to tomato and tobacco plants reduces the concentration of ToMV and TMV in tomato and bell pepper seedlings, respectively. RT-PCR products showed higher expression of two viral resistance genes viz., alternative oxidase (AOX) and RNA-dependent RNA polymerase (RdRp) in the upper leaves of the ASM-treated tomato plants challenge inoculated with ToMV. Viral concentration in the upper leaves was also reduced indicating impaired viral movement to upper leaves (Madhusudhan *et al.*, 2008). A pyrazolecarboxylic acid derivative also induces SAR in tobacco against TMV, but does not require salicylic acid (Yasuda *et al.*, 2003). Bion-induced LAR and SAR to TNV in *Phaseolus vulgaris* (Faoro and Iriti, 2006), and was also found to protect tobacco plants against the effects of the thrips-vectored TSWV in field-grown tobacco (Csinos *et al.*, 2001). BTH applied as a drench also ISR in tomato to CMV (Afoka, 2000).

2,6-dichloroisonicotinic acid (INA) enhances resistance to TMV. INA induces the expression of SAR genes, sometimes before the challenging inoculation and, in other cases, after pathogen attack only (Conrath *et al.*, 1995). It acts independently of the presence of SA that is like Bion it is a functional analog of salicylic acid and activates defensive signal transduction downstream of salicylic acid (Gozzo, 2003).

Riboflavin- (vitamin B2) ISR in tobacco to TMV when the plants were inoculated 4–5 days after treatment with 0.5 mM riboflavin (Dong and Beer, 2000). Thiamin (vitamin B1) also induced SAR in plants. Thiamine-treated rice, *Arabidopsis*, and vegetable crop plants showed resistance to fungal, bacterial, and viral infections through the salicylic acid and Ca^2 related signaling pathways (Ahn *et al.*, 2005).

CAP-34, a 34-kDa basic protein isolated from *Clerodendrum aculeatum*, ISR against *Sunnhemp rosette virus* in *Cyamopsis tetragonoloba*. Following treatment of *Cyamopsis* with CAP-34, 1,3- β-glucanase activity, an enzyme that degrades polysaccharidic substrates, increased rapidly. This enzyme has been implicated in SAR through release of endogenous elicitor molecules (Vivek *et al.*, 2001).

External application of purified beetin (a virus-inducible type 1 ribosome-inactivating protein from sugar beet) to sugar beet leaves prevented infection by *Artichoke mottled crinkle virus* (AMCV). This supports the hypothesis that beetins could be involved in plant SAR (Iglesias *et al.*, 2005).

A chemical inducer of SAR was isolated from the extracts of *Strobilanthes cusia* and identified to be 3-acetonyl-3-hydroxyoxindole (AHO), a derivative of isatin. Tobacco plants treated with AHO exhibited enhanced resistance to TMV, with accumulation of salicylic acid (SA) and phenylalanine ammonia-lyase activity (Li *et al.*, 2008).

Treatment of plants with non-toxic concentrations of ions of the heavy metal cadmium induces resistance to plant viruses (Ghoshroy *et al.*, 1998). Ueki and Citovsky (2002) identified a glycine-rich protein that promotes the accumulation of callose in the vascular tissue. This callose build up might restrict the unloading of viruses out of the phloem and inhibits the systemic virus movement.

Apparently, the mechanism that inhibits virus replication in the local lesion area and activates induced resistance can be turned on by various inducing compounds.

III. CONCLUDING THOUGHTS

Major advances in understanding the mechanism of localization and induced resistance have been made since reviewing this subject 30 years ago (Loebenstein, 1972). At that time it became evident that necrotization and ultrastructural changes around the lesion were not the major cause for restricting the virus within the lesion. These predictions that necrotization and localization are two separate processes got support from additional experimental data, as with PVX in Rx potato and TCV in resistant *Arabidopsis* (Kachroo *et al.*, 2000; Köhm *et al.*, 1993). Since then efforts were made to link PR proteins as the main factor in the localization of the virus, but so far no evidence has been provided that they are active in localizing the virus. They are not induced in virus infections that do not cause necrosis and may therefore be a host response to necrosis (or other stresses). No antiviral activity of any of the PR-proteins has so far been reported.

A major achievement was in the early 1990s, when the *N* gene was isolated by transposon tagging (Dinesh-Kumar *et al.*, 1995; Whitham *et al.*, 1994) and its transcription strategy determined. The *N* gene encodes a protein of 131.4 kDa, which has three domains: an N-terminal domain similar to that of the *Drosophila* Toll protein and the interleukin-1 receptor (TIR) in mammals, an NBS, and four imperfect LRR. The *N* gene thus belongs to the TIR-NBS-LRR class of *R* genes. It may be that the *N* protein activates the HR but activation of resistance may require another gene, similar to the *HRT* and *RRT* genes in TCP resistant *Arabidopsis* (Kachroo *et al.*, 2000).

The N protein may also function as a receptor that interacts with the gene product of TMV that elicits HR, perhaps by activating a transcription factor that induces the expression of genes responsible for the HR (Whitham *et al.*, 1994). It is suggested that the HR and inhibition of virus replication are two separate processes. It may also be that inhibition of viral RNA translation by is through RNA silencing (see Chapter 1). Thus in some cases, HR may be activated but fails to restrict virus multiplication or movement resulting in systemic movement.

The N protein apparently does not inhibit virus replication by itself but induces a cascade of events. We speculate that IVR or an IVR-like protein is produced at the end of the cascade. This protein(s) could be the main factor responsible for inhibiting virus replication resulting in localizing the infection and subsequently in SAR. The IVR protein not only inhibited TMV, but also PVX, PVY, and CMV in leaf disks floated on an IVR solution, and may perhaps be a broad-spectrum interferon-like inhibitor.

As to the mode of action of IVR, a speculative possibility is that it binds small RNAs (small interfering RNA or micro-RNA), which are involved in transcript turnover, cleavage, and translational control (Hutvagner and Zamore, 2002), and degrade viral RNA [see Csorba *et al.* (2009)] in this Volume]. It was shown that in several plants small RNA binding proteins were found in the phloem, which bound selectively 25-nt single-stranded RNA species (Yoo *et al.*, 2004). These proteins were in the range of 20.5–27 kDa, while IVR was about 23 kDa. These proteins could mediate the cell-to-cell trafficking of the siRNA's.

Plants transformed with NC330, encoding an IVR-like protein, resulted in a number of transgenic plants expressing variable resistance to TMV and the fungal pathogen *Botrytis cinerea* (Akad *et al.*, 2005). It will be interesting to see if transformation of different plants with sequences encoding IVR-like proteins will induce resistance to different viruses and perhaps to other pathogens. It might be that resistance of plants to viruses and fungi is associated with siRNA, which recently became known as a more general gene silencing agent, and may be associated with resistance in a variety of organisms.

The induced resistance phenomenon has attracted researchers over the last 50 years. The fact that many plants respond to infection by different pathogens—viruses, fungi, bacteria and even damage by insects, and various chemical compounds, in developing a certain degree of resistance in other non-infected parts may imply that this mechanism is of primeval origin. Its mechanism however is not clear. Suggestions have been made *inter alia* with its resemblance of the interferon mechanism in animals, association with small RNAs, PR proteins etc.

The findings that various natural and synthetic compounds activate induced resistance have raised hopes that this may become another tool in the strategy of crop protection. As salicylic acid seems to be involved in development of SAR, ASM also called BION, a synthetic analogue of salicylic acid was found to protect cantaloupe against CMV and reduced the concentration of ToMV and TMV in tomato and bell pepper seedlings, respectively. A novel strategy for control of *African cassava mosaic virus* (ACMV by mimicking a hypersensitive reaction using *barnase* (the ribunuclease produced by *Bacillus amyloliquefaciens*) seems also of interest. So far none of these compounds are in large scale practical uses. It is however the hope that future research will reach the field.

REFERENCES

Abbink, T. E. M., Tjernberg, P. A., Bol, J. F., and Linthorst, H. J. M. (1998). Tobacco mosaic virus helicase domain induces necrosis in *N* gene-carrying tobacco in the absence of virus replication. *Mol. Plant-Microbe Interact.* **11:**1242–1246.

Afoka, G. H. (2000). Benzo-(1, 2, 3)-thiadiazole-7-carbothioic acid S-methyl ester induces systemic resistance in tomato (*Lycopersicon esculentum* Mill cv. Vollendung) to Cucumber mosaic virus. *Crop Protect.* **19:**401–405.

Ahl, P., and Gianinazzi, S. (1982). b-Protein as a constitutive component in highly (TMV) resistant interspecific hybrids of *Nicotiana glutinosa* x *Nicotiana debneyi. Plant Sci. Letters* **26:**173–181.

Ahn, I. P., Kim., S. N., and Lee, Y. H. (2005). Vitamin B_1 functions as an activator of plant disease resistance. *Plant Physiol.* **138:**1505–1515.

Akad, F., Teverovsky, E., David, A., Czosnek, H., Gidoni, D., Gera, A., and Loebenstein, G. (1999). A cDNA from tobacco codes for an inhibitor of virus replication (IVR)- like protein. *Plant Mol. Biol.* **40:**969–976.

Akad, F., Teverovsky, E., Gidoni, D., Elad, Y., Kirshner, B., Rav-David, D., Czosnek, H., and Loebenstein, G. (2005). Resistance to *Tobacco mosaic virus* and *Botrytis cinerea* in tobacco transformed with the NC330 cDNA encoding an inhibitor of viral replication (IVR)-like protein. *Ann. Appl. Biol.* **147:**89–100.

Allan, A. C., Lapidot, M., Culver, J. N., and Fluhr, R. (2001). An early tobacco mosaic virus-induced oxidative burst in tobacco indicates extracellular perception of the virus coat protein. *Plant Physiol.* **126:**97–108.

Arvind, L., Dixit, V. M., and Koonin, E. V. (1999). The domains of death: Evolution of the apoptosis machinery. *Trends Biochem. Sci.* **24:**47–53.

Attaran, E., Zeer, T. E., Griebel, T., and Zeier, J. (2009). Methyl salicylate production and jasmonate signalling are not essential for systemic acquired resistance in *Arabidopsis. Plant Cell* **21:**954–961.

Azevedo, C., Sadanandom, A., Kitagawa, K., Freialdenhoven, A., Shirasu, K., and Schulze-Lefert, P. (2002). The RAR1 interactor SGT1, an essential component of R gene-triggered disease resistance. *Science* **295:**2073–2076.

Bendahmane, A., Kanyuka, K., and Baulcombe, D. C. (1999). The *Rx* gene from potato controls separate virus resistance and cell death responses. *Plant Cell* **11:**781–791.

Bent, A. F., Kunkel, B. N., Dahlbeck, D., Brown, K. L., Schmidt, R., Giraudat, J., Leung, J., and Staskawicz, B. J. (1994). RPS2 of *Arabidopsis thaliana*: A leucine-rich repeat class of plant disease resistance genes. *Science* **265:**1856–1860.

Boukema, I. W. (1980). Allelism of genes controlling resistance to TMV in *Capsicum* **L**. *Euphytica* **29:**433–439.

Carr, J.P., Lewsey, M.G., and Palukaitis, P. (in press). Signaling in induced resistance. *Adv. Virus Res.*

Canto, T., and Palukaitis, P. (1999). The hypersensitive response to cucumber mosaic virus in *Chenopodium amaranticolor* requires virus movement outside the initially infected cell. *Virology* **265:**74–82.

Canto, T., and Palukaitis, P. (2002). Novel N-gene associated, temperature-independent resistance to the movement of *Tobacco mosaic virus* vectors neutralized by a *Cucumber mosaic virus* RNA1 transgene. *J. Virol.* **76:**12908–12916.

Cawly, J., Cole, A. B., Király, L., Qiu, W., and Schoelz, J. E. (2005). The plant gene *CCD1* selectively blocks cell death during the hypersensitive response to *Cauliflower mosaic virus* infection. *Mol. Plant-Microbe Interact.* **18:**212–219.

Cheo, P. C. (1970). Subliminal infection of cotton by tobacco mosaic virus. *Phytopathology* **60:**41–46.

Chivasa, S., and Carr, J. P. (1998). Cyanide restores *N* gene-mediated resistance to tobacco mosaic virus in transgenic tobacco expressing salicylic acid hydroxylase. *Plant Cell* **10:**1489–1498.

Chivasa, S., Murphy, A.. M.., Naylor, M., and Carr, J. P. (1997). Salicylic acid interferes with tobacco mosaic virus replication via a novel salicylhydroxamic acid–sensitive mechanism. *Plant Cell* **9:**547–557.

Chu, M. H., Park, J. W., and Scholthof, H. B. (1999). Separate regions on the tomato bushy stunt virus p22 protein mediate cell-to-cell movement versus elicitation of effective resistance responses. *Mol. Plant-Microbe Interact.* **12:**285–292.

Clausen, R. E., and Goodspeed, T. H. (1925). Interspecific hybridization in *Nicotiana*. II. A tetraploid *glutinosa-tabacum* hybrid, an experimental verification of Winge's hypothesis. *Genetics* **10:**278–284.

Clarke, J. D., Volko, S. M., Ledford, H., Ausubel, F. M., and Dong, X. (2000). Roles of salicylic acid, jasmonic acid, and ethylene in *cpr*-induced resistance in *Arabidopsis*. *Plant Cell* **12:**2175–2190.

Cockerham, G. (1955). Strains of potato virus X. *Proc. 2nd Conf. Potato Virus Diseases*. Lisse-Wageningen 1954, pp. 89-92.

Cohen, J., and Loebenstein, G. (1975). An electron microscope study of starch lesions in cucumber cotyledons infected with tobacco mosaic virus. *Phytopathology* **65:**32–39.

Cole, A. B., Kiraly, L., Ross, K., and Schoelz, J. E. (2001). Uncoupling resistance from cell death in the hypersensitive response of *Nicotiana* species to *Cauliflower mosaic virus* infection. *Mol. Plant-Microbe Interact.* **14:**31–41.

Cooley, M. B., Pathirana, S., Wu, H. J., Kachroo, P., and Klessig, D. F. (2000). Members of the *Arabidopsis* HRT/RPP8 family of resistance genes confer resistance to both viral and oomycete pathogens. *Plant Cell* **12:**663–676.

Conrath, U., Chen, Z., Ricigliano, J. R., and Klessig, D. F. (1995). Two inducers of plant defense responses, 2,6-dichloroisonicotinic acid and salicylic acid, inhibit catalase activity in tobacco. *Proc. Natl. Acad. Sci. U. S. A.* **92:**7143–7147.

Cordelier, S., Ruffray, P., de, Fritig, B., and Kauffmann, S. (2003). Biological and molecular comparison between localized and systemic acquired resistance induced in tobacco by a *Phytophthora megasperma* glycoprotein elicitin. *Plant Mol. Biol.* **51:**109–118.

Csinos, A. S., Pappu, H. R., McPherson, R. M., and Stephenson, M. G. (2001). Management of *Tomato spotted wilt virus* in flue-cured tobacco with acibenzolar-S-methyl and imidacloprid. *Plant Dis.* **85:**292–296.

Csorba, T., Pantaleo, V., and Butgyan, J. (2009). RNA silencing, an antiviral mechanism. *Adv. Virus Res. (This volume).*

Culver, J. N. (2002). Tobacco mosaic virus assembly and disassembly: Determinants in pathogenicity and resistance. *Ann. Rev. Phytopathol.* **40:**287–308.

Culver, J. N., and Dawson, W. O. (1989). Tobacco mosaic virus coat protein: An elicitor of the hypersensitive reactions but not required for the development of mosaic symptoms in *Nicotiana sylvestris*. *Virology* **173:**755–758.

Culver, J. N., and Dawson, W. O. (1991). Tobacco mosaic virus elicitor coat protein genes produce a hypersensitive phenotype in transgenic *Nicotiana sylvestris* plants. *Mol. Plant-Microbe Interact.* **4:**458–463.

Culver, J. N., Lindebeck, A. G. C., and Dawson, W. O. (1991). Virus-host interaction: Induction of chlorotic and necrotic responses in plants by tobamoviruses. *Annu. Rev. Phytopathol.* **29:**193–217.

Culver, J. N., Stubbs, G., and Dawson, W. O. (1994). Structure-function relationship between tobacco mosaic virus coat protein and hypersensitivity in *Nicotiana sylvestris*. *J. Mol. Biol.* **242:**130–138.

Darby, R. M., Maddison, A., Mur, L. A. J., Bi, Y-M., and Draper, J. (2000). Cell-specific expression of salicylate hydrolase in an attempt to separate localized HR and systemic signalling establishing SAR in tobacco. *Mol Plant. Pathol.* **1:**115–123.

Dempsey, D. A., Pathirana, M. S., Wobbe, K. K., and Klessig, D. F. (1997). Identification of an *Arabidopsis* locus required for resistance to turnip crinkle virus. *Plant J.* **11:**301–311.

Dempsey, D. A., Wobbe, K. K., and Klessig, D. F. (1993). Resistance and susceptible responses of *Arabidopsis thaliana* to turnip crinkle virus. *Phytopathology* **83:**1021–1029.

Deom, C. M., Wolf, S., Holt, C. A., Lucas, W. J., and Beachy, R. N. (1991). Altered function of the tobacco mosaic virus movement protein in a hypersensitive host. *Virology* **180:**251–256.

Der, S. D., Zhou, A., Williams, B. R., and Silverman, R. H. (1998). Identification of genes differentially regulated by interferon alpha, beta, or gamma using oligonucleotide arrays. *Proc. Natl. Acad. Sci. U. S. A.* **95:**15623–15628.

De Jong, W., Forsyth, A., Leister, D., Gebhardt, C., and Baulcombe, D. C. (1997). A potato hypersensitive resistance gene against potato virus X maps to a resistance cluster on chromosome V. *Theor. Appl. Genet.* **95:**246–252.

Dietrich, R. A., Delaney, T. P., Uknes, S. J., Ward, E. R., Ryals, J. A., and Dangl, J. L. (1994). Arabidopsis mutants simulating disease resistance response. *Cell* **77:**565–577.

Dinesh-Kumar, S. P., Whitham, S., Choi, D., Hehl, R., Corr, C., and Baker, B. (1995). Transposon tagging of tobacco mosaic virus resistance gene N: Its possible role in the TMV-N-mediated signal transduction pathway. *Proc. Natl. Acad. Sci. U. S. A.* **92:**4175–4180.

Dinesh-Kumar, S. P., and Baker, B. J. (2000). Alternatively spliced *N* resistance gene transcripts: Their possible role in tobacco mosaic virus resistance. *Proc. Natl. Acad. Sci. U. S. A.* **97:**1908–1913.

Dinesh-Kumar, S. P., Wai-Hong Tham, W., and Baker, B. (2000). Structure-function analysis of the tobacco mosaic virus resistance gene *N*. *Proc. Natl. Acad. Sci. U. S. A.* **97:**14789–14794.

Dong, H., and Beer, S. V. (2000). Riboflavin induces disease resistance in plants by activating a novel signal transduction pathway. *Phytopathology* **90:**801–811.

Dunigan, D. D., Golemboski, D. B., and Zaitlin, M. (1987). Analysis of the *N gene of Nicotiana*. *In* "Plant Resistance to Viruses (Ciba Foundation Symposium 133)" (D. Evered and S. Harnett, eds.), pp. 120–135. John Wiley and Sons, Cichester.

Dunigan, D. D., and Madlener, J. C. (1995). Serine/threonine protein phosphatase is required for virus-mediated programmed cell death. *Virology* **207:**460–466.

Durner, J., and Klessig, D. F. (1995). Inhibition of ascorbate peroxidase by salicylic acid and 2,6-dichloroisonicotinic acid, two inducers of plant defense responses. *Proc. Natl. Acad. Sci. U. S. A.* **92:**11312–11316.

Ehrenfeld, N., Cañón, P., Stange, C., Medina, C., and Arce-Johnsons, P. (2005). Tobamovirus coat protein CPCg iInduces an HR-like response in sensitive tobacco plant. *Mol. Cells* **19:**418–427.

Ehrenfeld, N., Gonzalez, A., Canon, P., Medina, C., Perez-Acle, T., and Arce-Johnson, P. (2008). Structure-function relationship between the tobamovirus TMV-Cg coat protein and the HR-like response. *J. Gen. Virol.* **89:**809–881.

Elvira, M. I., Galdeano, M. M., Gilardi, P., Garcia-Luque, I., and Serra, M. T. (2008). Proteomic analysis of pathogenesis-related proteins (PRs) induced by compatible and incompatible interactions of pepper mild mottle virus (PMMoV) in *Capsicum chinense* L^3 plants. *J. Exp. Bot.* **59:**1253–1265.

Faoro, F., and Iriti, M. (2006). Cell death or not cell death: two different mechanisms for chitosan and BTH antiviral activity. *Bull.OILB/SROP* **29:**25–29.

Faoro, F., Sant, S., Iriti, M., Maffi, D., and Appiano, A. (2001). Chitosan-elicited resistance to plant viruses: a histochemical and cytohemical study. *In* "Chitin Enzymology;" (R. A. A. Muzzarelli, ed.), pp. 57–62. AtecEdizioni, Grottammare, Italy.

Faulkner, C., and Kimmins, W. C. (1978). Staining reactions of the tissue bordering lesions induced by wounding and virus infections. *Can. J. Bot.* **56:**2980–2999.

Favali, M. A., Bassi, M., and Conti, G. G. (1974). Morphological cytochemical and autoradiographic studies of local lesions induced by the U5 strain of tobacco mosaic virus in *Nicotiana glutinosa* L. *Riev. Patolog. Vegetale Series IV* **10:**207–218.

Fodor, J., Hideg, E., Kecskes, A., and Kiraly, Z. (2001). In vivo detection of tobacco mosaic virus-induced local and systemic oxidative burst by electron paramagnetic resonance spectroscopy. *Plant Cell Physiol.* **42:**775–779.

Genda, Y., Kanda, A., Hamada, H., Sato, K., Ohnishi, J., and Tsuda, S. (2007). Two amino acid substitutions in the coat protein of *Pepper mild mottle virus* are responsible for overcoming the L^4 gene-mediated resistance in *Capsicum* spp. *Phytopathology* **97:**787–793.

Gera, A., and Loebenstein, G. (1983). Further studies of an inhibitor of virus replication from tobacco mosaic virus-infected protoplasts of a local lesion-responding tobacco cultivar. *Phytopathology* **73:**111–115.

Gera, A., and Loebenstein, G. (1988). An inhibitor of virus replication associated with green island tissue of tobacco infected with cucumber mosaic virus. *Physiol Mol. Plant Pathol.* **32:**373–385.

Gera, A., Loebenstein, G., and Shabtai, S. (1983). Enhanced tobacco mosaic virus production and suppressed synthesis of a virus inhibitor in protoplasts exposed to antibiotics. *Virology* **127:**475–478.

Gera, A., Loebenstein, G., Salomon, R., and Franck, A. (1990). An inhibitor of virus replication (IVR) from protoplasts of a hypersensitive tobacco cultivar infected with tobacco mosaic virus is associated with a 23K protein species. *Phytopathology* **80:**78–81.

Gera, A., Spiegel, S., and Loebenstein, G. (1986). Production, preparation and assay of an antiviral substance from plant cells. *Methods Enzymol.* **119:**729–733.

Gera, A., Tam, Y., Teverovsky, E., and Loebenstein, G. (1993). Enhanced tobacco mosaic virus production and suppressed synthesis of the inhibitor of virus replication in protoplasts and plants of local lesion responding cultivars exposed to 35°C. *Physiol. Mol. Plant Pathol.* **43:**299–306.

Gera, A., Tostado, J. M., Garcia-Luque, I., and Serra, M. T. (1994). Association of the inhibitor of virus replication with resistance mechanisms in pepper cultivars. *Phytoparasitica* **22:**219–227.

Ghoshroy, S., Freedman, K., Lartey, R., and Citovsky, V. (1998). Inhibition of plant viral systemic infection by non-toxic concentrations of cadmium. *Plant J.* **13**:591–602.

Gianinazzi, S., and Kassanis, B. (1974). Virus resistance induced in plants by polyacrylic acid. *J. Gen. Virol.* **23:**1–9.

Gicherman, G., and Loebenstein, G. (1968). Competitive inhibition by foreign nucleic acids and induced interference by yeast-RNA with the infection of tobacco mosaic virus. *Phytopathology* **58:**405–409.

Gilardi, P., García-Luque, I., and Serra, M. T. (2004). The coat protein of tobamovirus acts as elicitor of both L2 and L4 gene-mediated resistance in *Capsicum. J. Gen. Virol.* **85:**2077–2085.

Gilliland, A., Murphy, A. M., and Carr, J. P. (2006). Induced resistance mechanisms. *In* "Natural Resistance Mechanisms" (G. Loebenstein and J. P. Carr, eds.), pp. 125–145. Springer, Dordrecht, The Netherlands.

Gilpatrick, J. D., and Weintraub, M. (1952). An unusual type of protection with the carnation mottle virus. *Science* **115**:101–102.

Goulden, M. G., and Baulcombe, D. C. (1993). Functionally homologous host components recognize potato virus X in *Gomphrena globosa* and potato. *Plant Cell* **5:**921–930.

Gozzo, F. (2003). Systemic acquired resistance in crop protection: From nature to a chemical approach. *J. Agric. Food Chem.* **5:**4487–4503.

Green, D. R., and Reed, J. C. (1998). Mitochondria and apoptosis. *Science* **281:**1309–1311.

Guo, A., Salih, G., and Klessig, D. F. (2000). Activation of a diverse set of genes during the tobacco resistance response to TMV is independent of salicylic acid; induction of a subset is also ethylene independent. *Plant J.* **21:**409–418.

Hamada, H., Tomita, R., Iwadate, Y., Kobayashi, K., Munemura, I., Takeuchi, S., Hikichi, Y., and Suzuki, K. (2007). Cooperative effect of two amino acid mutations in the coat protein of Pepper mild mottle virus overcomes L^3-mediated resistance in *Capsicum* plants. *Virus Genes* **34:**205–214.

Hamäläinen, J. H., Sorri, V. A., Watanabe, K. N., Gebhardt, C., and Valkonen, J. P. T. (1998). Molecular examination of a chromosome region that controls resistance to potato Y and A potyviruses in potato. *Theor. Appl. Genet.* **96:**1036–1043.

Harrison, B.D. (1955). Studies on virus multiplication in inoculated leaves. *Ph.D. Thesis*, London University.

Heidel, A. J., Clarke, J. D., Antonovics, J., and Dong, X. (2004). Fitness costs of mutations affecting the systemic acquired resistance pathway in Arabidopsis thaliana. *Genetics* **168**:2197–2206.

Herbers, K., Meuwly, P., Frommer, W. B., Métraux, J. P., and Sonnewald, U. (1996). Systemic acquired resistance mediated by the ectopic expression of invertase: possible hexose sensing in the secretory pathway. *Plant Cell* **8:**793–803.

Hill, J. H. (2003). Soybean. *In* "Virus and Virus-like Diseases of Major Crops in Developing Countries" (Gad Loebenstein and George Thottappilly, eds.), pp. 377–395. Kluwer Academic Publishers, Dordrecht, The Netherlands.

Holmes, F. O. (1929). Local lesions in tobacco mosaic. *Bot. Gaz.* **87:**39–70.

Holmes, F. O. (1938). Inheritance of resistance to tobacco mosaic disease in tobacco. *Phytopathology* **28:**553–561.

Holmes, F. O. (1954). Inheritance of resistance to viral diseases in plants. *Adv. Virus Res.* **2:**1–30.

Honda, A., Takahashi, H., Toguri, T., Ogawa, T., Hase, S., Ikegami, M., and Ehara, Y. (2003). Activation of defense-related gene expression and systemic acquired resistance in cucumber mosaic virus-infected tobacco plants expressing the mammalian 2′5′ oligoadenylate system. *Arch. Virol.* **148:**1017–1026.

Hu, G. S., Hart, A. K. A., Li, Y. S., Ustach, C., Handley, V., Navarre, R., Hwang, C. F., Aegerter, B. J., Williamson, V. M., and Baker, B. (2005). EDS1 in tomato is required for resistance mediated by TIR-class R genes and the receptor-like R gene Ve. *Plant J.* **42:**376–391.

Hubert, D., Tornero, P., Belkhadir, Y, Krishna, P., Takahashi, A., Shirasu, K., and Dangl, J.. L (2003). Cytosolic HSP90 associates with and modulates the *Arabidopsis* RPM1 disease resistance protein. *EMBO J.* **22:**5679–5689.

Huang, W. E, Huang, L. F., Preston, G. M., Naylor, M., Carr, J. P., Li, Y. H., Singer, A. C., Whiteley, A. S., and Wang, H. (2006). Quantitative in situ assay of salicylic acid in tobacco leaves using a genetically modified biosensor strain of *Acinetobacter* sp. ADP1. *Plant J.* **46:**1073–1083.

Huffaker, A., Pearce, G., and Ryan, C. A. (2006). An endogenous peptide signal in *Arabidopsis* activates components of the innate immune response. *Proc. Natl. Acad. Sci. U. S. A.* **103:**10098–10103.

Hutvagner, G., and Zamore, P. D. (2002). A microRNA in a multiple-turnover RNAi enzyme complex. *Science* **297:**2056–2060.

Iglesias, R., Perez, Y., Torre, C., de, Ferreras, J. M., Antolin, P., Jimenez, P., Rojo, M. A., Mendez, E., and Girbes, T. (2005). Molecular characterization and systemic induction of single-chain ribosome-inactivating proteins (RIPs) in sugar beet (*Beta vulgaris*) leaves. *J. Exp. Bot.* **56:**1675–1684.

Iriti, M., Sironi, M., Gomarasca, S., Casazza, A. P., Soave, C., and Faoro, F. (2006). Cell death-mediated antiviral effect of chitosan in tobacco. *Plant Physiol. Biochem.* **44:**893–900.

Ji, L. H. and Ding, S. W. (2001). The suppressor of transgene RNA silencing encoded by *Cucumber mosaic virus* interferes with salicylic acid-mediated virus resistance. *Mol. Plant-Microbe Interact.* **14**:715–724.

Jones, R. A. C. (1990). Strain group specific and virus specific hypersensitive reactions to infection with potyviruses in potato. *Ann. Appl. Biol.* **117:**93–105.

Kachroo, P., Yoshioka, K., Shah, J., Dooner, H,K, and Klessig, D. F. (2000). Resistance to turnip crinkle virus in *Arabidopsis* is regulated by two host genes and is salicylic acid dependent but NPR1, ethylene, and jasmonate independent. *Plant Cell* **12:**677–690.

Kang, B.-C., Yeam, I., and Jahn, M. M. (2005). Genetics of plant virus resistance. *Ann. Rev. Phytopathol.* **43:**581–621.

Karger, E. M., Frolova, O. Yu., Fedorova, N. V., Baratova, L. A., Ovchinnikova, T. V., Susi, P., Makinen, K., Ronnstrand, L., Dorokhov, Yu. L, and Atabekov, J. G (2003). Dysfunctionality of a tobacco mosaic virus movement protein mutant mimicking threonine 104 phosphorylation. *J. Gen. Virol.* **84:**727–732.

Kauffmann, S., Legrand, M., Geoffery, P., and Frittig, B. (1987). Biological function of pathogenesis-related' proteins: four PR proteins of tobacco have 1,3-β-glucanase activity. *EMBO J.* **6:**3209–3212.

Kavanagh, T., Goulden, M., Santa Cruz, S., Chapman, S., Barker, I., and Baulcombe, D. (1992). Molecular analysis of a resistance-breaking strain of potato virus X. *Virology* **189:**609–617.

Kim, C.-H., and Palukaitis, P. (1997). The plant defense response to cucumber mosaic virus in cowpea is elicited by the viral polymerase gene and affects virus accumulation in single cells. *EMBO J.* **16:**4060–4068.

Kiraly, Z., Barna, B.., Kecskes, A., and Fodor, J. (2002). Down-regulation of antioxidative capacity in a transgenic tobacco which fails to develop acquired resistance to necrotization caused by TMV. *Free Radical Res.* **36:**981–991.

Király, Z., and Király, L. (2006). To die or not to die - Is cell death dispensable for resistance during the plánt hypersensitive response. *Acta Phytopathol. Entomol. Hung.* **41:**11–21.

Kitajima, E. W., and Costa, A. S. (1973). Aggregates of chloroplasts in local lesions induced in *Chenopodium quinoa* Wild. by turnip mosaic virus. *J. Gen. Virol.* **20:**413–416.

Kitajima, E. W, Rezende, J. A, and Rodrigues, J. C. (2003). Passion fruit green spot virus vectored by *Brevipalpus phoenicis* (Acari: Tenuipalpidae) on passion fruit in Brazil. *Exp Appl Acarol.* **30:**225–231.

Klarzynsky, O., Descamps, V., Plesse, B., Yvin, J-C., Kloareg, B., and Fritig, B. (2003). Sulfated fucan oligosaccharides elicit defebse responses in tobacco and local an systemic resistance against Tobacco mosaic virus. *Mol. Plant-Microbe Interact.* **16:**115–122.

Kleczkowsky, A. (1950). Interpreting relationships between concentrations of plant viruses and numbers of local lesions. *J. Gen. Microbiol.* **4:**53–69.

Knorr, D. A., and Dawson, W. O. (1988). A point mutation in the tobacco mosaic virus capsid protein gene induces hypersensitivity in Nicotiana sylvestris. *Proc. Natl. Acad. Sci. U. S. A.* **85:**170–174.

Köhm, B. A., Goulden, M. G., Gilbert, J. E., Kavanagh, T. A., and Baulcombe, D. C. (1993). A potato virus X resistance gene mediates an induced, nonspecific resistance in protoplasts. *Plant Cell* **5:**913–920.

Kovalenko, A. G., Grabina, T. D., Kolesnik, L. V., Didenko, L. F., Oleschenko, L. T., Olevinskaya, Z. M., and Telegaeva, T. A. (1993). Virus resistance induced by mannan sulphates in hypersensitive host plants. *J. Phytopathol.* **137:**133–147.

Krasavina, M. S., Malyshenko, S. I., Raldugina, G. N., Burmistrova, N. A., and Nosov, A. V. (2002). Can salicylic acid affect the intercellular transport of the Tobacco mosaic virus by changing plasmodesmal permeability? *Russian J. Plant Physiol.* **49:**61–67.

Kubo, S., Ikeda, T., Takanami, Y., and Mikami, Y. (1990). A potent plant virus inhibitor found in *Mirabilis jalapa* L. *Ana. Phytopathol. Soc. Japan* **56:**481–487.

Kumar, D., Gustafsson, C, and Klessig, D. F. (2006). Validation of RNAi silencing specificity using synthetic genes: salicylic acid-binding protein 2 is required for innate immunity in plants. *Plant J.* **45:**863–868.

Kumar, D., and Klessig, D. F. (2003). High-affinity salicylic acid-binding protein 2 is required for plant innate immunity and has salicylic acid-stimulated lipase activity. *Proc. Natl. Acad. Sci. U. S. A.* **100:**16101–16106.

Laat, A. M. M. de, and Loon, L. C van (1983). The relationship between stimulated ethylene production and symptom expression in virus-infected tobacco leaves. *Physiol. Plant Pathol.* **22:**261–273.

Lacomme, C., and Santa Cruz, S. (1999). Bax-induced cell death in tobacco is similar to the hypersensitive response. *Proc. Natl. Acad. Sci.* **96:**7956–7961.

Legrand, M., Kauffmann, S., Geoffrey, P., and Frittig, B. (1987). Biological function of pathogenesis-related proteins: four tobacco pathogenesis-related proteins are chitinases. *Proc. Natl. Acad. Sci. U. S. A.* **84:**6750–6754.

Leonard, D. A., and Zaitlin, M. (1982). A temperature sensitive strain of tobacco mosaic virus defective in cell-to-cell movement generates an altered viral-coded protein. *Virology* **117:**416–424.

Li, X. H., and Simon, A. E. (1990). Symptom intensification on cruciferous hosts by the virulent sat-RNA of turnip crinkle virus. *Phytopathology* **80:**238–242.

Li, Y., Zhang, Z., Jia, Y., Shen, Y., He, H., Fang, R., Chen, X., and Hao, X. (2008). 3-Acetonyl-3-hydroxyoxindole: a new inducer of systemic acquired resistance in plants. *Plant Biotechn. J.* **6**:301–308.

Linthorst, H. J. (1991). Pathogenesis-related proteins of plants. *Crit. Rev. Plant Sci.* **10:**123–150.

Liu, Y., Jin, H., Yang, K., Kim, C. A., Baker, B., and Zhang, S. (2003). Interaction between two mitogen-activated protein kinases during tobacco defense signalling. *Plant J.* **34:**149–160.

Liu, Y., Schiff, M., Marathe, R., and Dinesh-Kumar, S. P. (2002). Tobacco *Rar1*, *EDS1* and *NPR1/NIM1* like genes are required for *N*-mediated resistance to tobacco mosaic virus. *Plant J.* **30:**415–429.

Liu, Y., Schiff, M., Serino, G., Deng, X-W., and Dinesh-Kumar, S. P. (2002). Role of SCF ubiquitin—ligase and the COP9 signalosome on the N gene-mediated resistance response to *Tobacco mosaic virus*. *Plant Cell* **14:**1483–1496.

Liu-Chun, Y., Chen-Qing, S., Xin-Da, W., Qiu-Hong, M., and Shan-Da, P. (2005). EST analysis of resistance to soybean mosaic virus (SMV) in soybean at the primary infected stage. *Acta Agron. Sinica.* **31:**1394–1399.

Loebenstein, G. (1972). Localization and induced resistance in virus-infected plants. *Ann. Rev. Phytopathol.* **10:**177–206.

Loebenstein, G., Chazan, R., and Eisenberg, M. (1970). Partial suppression of the localizing mechanism to tobacco mosaic virus by UV irradiation. *Virology* **41:**373–376.

Loebenstein, G., and Gera, A. (1981). Inhibitor of virus replication released from tobacco mosaic virus-infected protoplasts of a local lesion-responding tobacco cultivar. *Virology* **114**:132–139.

Loebenstein, G., Gera, A., Barnett, A., Shabtai, S., and Cohen, J. (1980). Effect of 2,4-dichlorophenoxyacetic acid on multiplication of tobacco mosaic virus in protoplasts from local-lesion and systemic-responding tobaccos. *Virology* **100:**110–115.

Loebenstein, G., Gera, A, and Gianinazzi, S. (1990). Constitutive production of an inhibitor of virus replication in the interspecific hybrid of *Nicotiana glutinisa* X *Nicotiana debneyi*. *Physiol. Plant Pathol.* **37:**145–151.

Loebenstein, G., and Lovrekovich, L. (1966). Interference with tobacco mosaic virus local lesion formation in tobacco by injecting heat-killed cells of *Pseudomonas syringae*. *Virology* **30:**587–591.

Loebenstein, G., Sela, B., and Van Praagh, T. (1969). Increase of tobacco mosaic local lesion size and virus multiplication in hypersensitive hosts in the presence of actinomycin D. *Virology* **37:**42–48.

Loebenstein, G., Rav David, D., Leibman. D., Gal-On, A., RonVunsh, R., Czosnek, H., and Elad, Y. (in press). Tomato plants transformed with the Inhibitor of Virus Replication (IVR) gene are partially resistant to *Botrytis cinerea*. *Phytopathology.*

Loebenstein, G., Spiegel, S., and Gera, A. (1982). Localized resistance and barrier substances. *In* "Active Defense Mechanisms in Plants" (R. K. S. Wood, ed.), pp. 211–230. Plenum Press, New York.

Lovato, F. A., Inoue-Nagata, A. K., Nagata, T., Avila, A. C., de, Pereira, L. A. R., and Resende, R. O. (2008). The N protein of *Tomato spotted wilt virus* (TSWV) is associated with the induction of programmed cell death (PCD) in *Capsicum chinense* plants, a hypersensitive host to TSWV infection. *Virus Res.* **137:**245–252.

Madhusudhan, K. N., Deepak, S. A., Prakash, H. S., Agrawal, G. K., Jwa, N. S., and Rakwal, R. (2008). Acibenzolar-S-methyl (ASM)-induced resistance against tobamoviruses involves induction of RNA-dependent RNA polymerase (RdRp) and alternative oxidase (AOX) genes. *J. Crop Sci. Biotech.* **11:**127–134.

Malamy, J., Carr, J. P., Klessig, D. F., and Raskin, I. (1990). Salicylic acid: a likely endogenous signal in the resistance response of tobacco to viral infection. *Science* **250:**1002–1004.

Maldonado, A. M., Doerner, P., Dixon, R. A., Lamb, C. J., and Cameron, R. K. (2002). A putative lipid transfer protein involved in systemic resistance signalling in *Arabidopsis*. *Nature* **419:**399–403.
Mandal, B., Mandal, S., Csinos, A. S., Martinez, N., Culbreath, A. K., and Pappu, H. R. (2008). Biological and molecular analyses of the acibenzolar-S-methyl-induced systemic acquired resistance in flue-cured tobacco against *Tomato spotted wilt virus*. *Phytopathology* **98:**196–204.
Mao, J. J., Qiu, D. W., Yang, X. F., Zeng, H. M., and Yuan, J. J. (2008). Expression of protein elicitor-encoding gene *pemG1* in tobacco (*Nicotiana tobacum* cv. Samsun NN) plants and enhancement of resistance to TMV. *Acta Agron. Sinica.* **34:**2070–2076.
Marathe, R., Anandalakshmi, R., Liu, Y., and Dinesh-Kumar, S. P. (2002). The tobacco mosaic resistance gene, *N*. *Mol. Plant Pathol.* **3:**167–172.
Menard, R., Ruffray, P., de, Fritig, B., Yvin, J. C., and Kauffmann, S. (2005). Defense and resistance-inducing activities in tobacco of the sulfated beta -1,3 glucan PS3 and its synergistic activities with the unsulfated molecule. *Plant Cell Physiol.* **46:**1964–1972.
Milne, R. C. (1966). Electron microscopy of tobacco mosaic virus in leaves of *Nicotiana glutinosa*. *Virology* **28:**527–532.
Morana, S. J., Wolf, C. M., Li, J., Reynolds, J. E., Brown, M. K., and Eastman, A. (1996). The involvement of protein phosphatase in the activation of ICE/ CED-3 protease, intercellular acidification, DNA digestion, and apoptosis. *J. Biol. Chem.* **271:**18263–18271.
Moser, O., Gagey, M. J., Godefroy-Colburn, T., Stussi-Garaud, C., Ellwar-Tschurtz, M., and Nitschko, H. (1998). The fate of the transport protein of tobacco mosaic virus in systemic and hypersensitive tobacco hosts. *J. Gen. Virol.* **69:**1367–1378.
Mur, L. A. J., Bi, Y-M., Darby, R. M., Firek, S., and Draper, J. (1997). Compromising early salicylic acid accumulation delays the hypersensitive response and increases viral dispersion during lesion establishment in TMV-infected tobacco. *Plant J.* **12:**1113–1126.
Murphy, A. M., Gilliland, A., Wong, C. E., West, J., Singh, D. P., and Carr, J. P. (2001). Induced resistance to viruses. *Eur. J. Plant Pathol.* **107**:121–128.
Murphy, A. M., and Carr, J. P. (2002). Salicylic acid has cell-specific effects on *Tobacco mosaic virus* replication and cell-to-cell movement. *Plant Physiol.* **128:**552–563.
Nakashita, H., Yoshioka, K., Yasuda, M., Nitta, T., Arai, Y., Yoshida, S., and Yamaguchi, I. (2002). Probenazole induces systemic acquired resistance in tobacco through salicylic acid accumulation. *Physiol. Mol. Plant Pathol.* **61:**197–203.
Nie, X. Z. (2006). Salicylic acid suppresses Potato virus Y isolate N:O-induced symptoms in tobacco plants. *Phytopathology* **96:**255–263.
Niehl, A., Lacomme, C., Erban, A., Kopka, J., Kramer, U., and Fisahn, J. (2006). Systemic Potato virus X infection induces defence gene expression and accumulation of beta -phenylethylamine-alkaloids in potato. *Functional Plant Biol.* **33**:593–604.
Oostendorp, M., Kunz, W., Dietrich, B., and Staub, T. (2001). Induced disease resistance in plants by chemicals. *Eur. J. Plant Pathol.* **107:**19–28.
Otsuki, A., Shimomura, T., and Takebe, I. (1972). Tobacco mosaic virus multiplication and expression of the N-gene in necrotic responding tobacco varieties. *Virology* **50:**45–50.

Padgett, H. S., and Beachy, R. N. (1993). Analysis of a tobacco mosaic virus strain capable of overcoming N gene-mediated resistance. *Plant Cell* **5:**577–586.

Padgett, H. S., Watanabe, Y., and Beachy, R. N. (1997). Identification of the TMV replicase sequence that activates the *N* gene-mediated hypersensitive response. *Mol. Plant-Microbe Interact.* **10:**709–715.

Park, S. W., Kaimoyo, E., Kumar, D., Mosher, S., and Klessig, D. F. (2007). Methyl salicylate is a critical mobile signal for plant systemic acquired resistance. *Science* **318:**113–116.

Penuela, S., Danesh, D., and Young, N. D. (2002). Targeted isolation, sequence analysis, and physical mapping of nonTIR NBS-LRR genes in soybean. *Theor. Appl. Genet.* **104:**261–272.

Pozo, O. de, l, and Lam, R. (2003). Expression of the Baculovirus p^{35} protein in tobacco affects cell death progression and compromises N gene-mediated disease resistance response to *Tobacco mosaic virus. Mol. Plant-Microbe Interact.* **16:**485–494.

Ross, A. F. (1961a). Localized acquired resistance to plant virus infection in hypersensitive hosts. *Virology* **14:**329–339.

Ross, A. F. (1961b). Systemic acquired resistance induced by localized virus infections in plants. *Virology* **14**:340–358.

Salmeron, J. M., Barker, S. J., Carland, F. M., Mehta, A. Y., and Staskawicz, B. J. (1994). Tomato mutants altered in bacterial disease resistance provide evidence for a new locus controlling pathogen recognition. *Plant Cell* **6:**511–520.

Saraste, M., Sibbald, P. R, and Wittinghofer, A. (1990). The P-loop a common motif in ATP- and GTP-binding proteins. *Trends Neurosci.* **15:**430–434.

Sekine, K. T., Kawakami, S., Hase, S., Kubota, M., Ichinose, Y., Shah, J., Kang, H. G., Klessig, D. F., and Takahashi, H. (2008). High level expression of a virus resistance gene, *RCY1*, confers extreme resistance to *Cucumber mosaic virus* in *Arabidopsis thaliana. Mol. Plant-Microbe Interact.* **21:**1398–1407.

Sela, B., Loebenstein, G., and Van Praagh, T. (1969). Increase of tobacco mosaic virus multiplication and lesion size in hypersensitive hosts in the presence of chloramphenicol. *Virology* **39:**260–264.

Seo, S., Okamoto, M., Iwai, T., Iwano, M., Fukui, K., Isogai, A., Nakajima, N., and Ohashi, Y. (2000). Reduced levels of chloroplast FtsH protein in tobacco mosaic virus–infected tobacco leaves accelerate the hypersensitive reaction. *Plant Cell* **12:**917–932.

Seo, S., Seto, H., Koshino, H., Yoshida, S., and Ohashi, Y. (2003). A diterpene as an endogenous signal for the activation of defense responses to infection with *Tobacco mosaic virus* and wounding in tobacco. *Plant Cell* **15:**863–873.

Siegrist, J., Orober, M., and Buchenauer, H. (2000). β-Aminobutyric acid-induced enhancement of resistance in tobacco mosaic virus depends on the accumulation of salicylic acid. *Physiol. Mol. Plant Pathol.* **56:**95–106.

Silverman, F. P., Petracek, P.-D., Heiman, D –F., Fledderman, C. M., and Warrior, P. (2005). Salicylate activity. 3. Structure relationship to systemic acquired resistance. *J. Agric. Food Chem.* **53:**9775–9780.

Shang, W. J., Wu, Y. F., Shang, H. S., Zhao, Xi. M, and Du, Y. G. (2007). Inhibitory effect to TMV-CP gene expression in tobacco induced by chito-oligosaccharides. *Acta Phytopathol. Sinica* **37:**637–641.

Shimomura, T. A., and Dijkstra, J. (1975). The occurrence of callose during the process of local lesion formation. *Neth. J. Plant Pathol.* **81:**107–121.
Shirasu, K., Lahaye, T., Tan, M. W., Zhou, F. S., Azevedo, C., and Schulze-Lefert, P. (1999). A novel class of eukaryotic zinc-binding proteins is required for disease resistance signaling in barley and development in *C. elegans. Cell* **99:**355–366.
Shulaev, V, Silverman, P, and Raskin, I (1997). Airborne signalling by methyl salicylate in plant pathogen resistance. *Nature* **385:**718–721.
Simon, A. E., Li, X. H., Lew, J. E., Stange, R., Zhang, C., Polacco, M., and Carpenter, C. D. (1992). Susceptibility and resistance of *Arabidopsis thaliana* to turnip crinkle virus. *Mol. Plant-Microbe Interact.* **5:**496–503.
Simons, T. J., and Ross, A. F. (1965). Effect of 2,4D-ddichlorophenoxyacetic acid on size of tobacco mosaic virus lesions in hypersensitive tobacco. *Phytopathology* **55:**1076–1077 (abstract).
Smith-Becker, J., Keen, N. T, and Becker, J. O. (2003). Acibenzolar-S-methyl induces resistance to *Colletotrichum lagenarium* and cucumber mosaic virus in cantaloupe. *Crop Protection* **22:**769–774.
Song, F. M., and Goodman, R. M. (2001). Activity of nitric oxide is dependent on, but is partially required for function of, salicylic acid in the signaling pathway in tobacco systemic acquired resistance. *Mol. Plant-Microbe Interact.* **14:**1458–1462.
Spiegel, S., Gera, A., Salomon, R., Ahl, P., Harlap, S., and Loebenstein, G. (1989). Recovery of an inhibitor of virus replication from the intercellular fluid of hypersensitive tobacco infected with tobacco mosaic virus and from uninfected induced-resistant tissue. *Phytopathology* **79:**258–262.
Stahmann, M. A., and Gothoskar, S. S. (1958). The inhibition of the infectivity of tobacco mosaic virus by some synthetic and natural polyelectrolytes. *Phytopathology* **48:**362–365.
Stein, A., and Loebenstein, G. (1970). Induction of resistance to tobacco mosaic virus by poly I: Poly C in plants. *Nature* **226:**363–364.
Sticher, L., Mauch-Mani, B., and Métraux., J. P. (1997). Systemic acquired resistance. *Ann. Rev. Phytopathol.* **35:**235–270.
Stange, C., Matus, J. M., Elorza, A., and Arce-Johnson, P. (2004). Identification and characterization of a novel tobacco mosaic virus resistance N gene homologue in *Nicotiana tabacum* plants. *Functional Plant Biol.* **31:**149–158.
Stussi-Garaud, C., Garaud, J. R., Berna, A., and Godefroy-Colburn, T. (1987). In situ location of alfalfa mosaic virus non-structural protein in plant cell walls: correlation with virus transport. *J. Gen. Virol.* **68:**1779–1784.
Takahashi, A., Casais, C., Ichimura, K., and Shirasu, K. (2003). HSP90 interacts with RAR1 and SGT1 and is essential for RPS2-mediated disease resistance in *Arabidopsis. Proc. Natl. Acad. Sci. U. S. A.* **100:**11777–11782.
Takusari, H., and Takahashi, T. (1979). Studies on viral pathogenesis in host plants. IX. Effect of citrinin on the formation of necrotic lesion and virus localization in the leaves of 'Samsun NN' tobacco plants after tobacco mosaic virus infection. *Phytopathol. Z* **96:**324–329.
Tobias, I., Rast, A, Th., B., and Maat, D. Z. (1982). Tobamoviruses of pepper, eggplant and tobacco: comparative host reactions and serological relationships. *Net. J. Plant Pathol.* **88:**257–268.

Tommenius, K., Clapham, D., and Meshi, T. (1987). Localization by immunogold cytochemistry of the virus-coded 30K protein in plasmodesmata of leaves infected with tobacco mosaic virus. *Virology* **160:**363–371.

Tommiska, T. J., Hamäläläinen, J. H., Watanabe, K. N., and Valkonen, J. P. T. (1998). Mapping of the gene Nx_{phu} that controls hypersensitive resistance to potato virus X in *Solanum phureja* lvP35. *Theor. Appl. Genet.* **9:**840–843.

Traw, M. B., Kim, J., Enright, S., Cipollini, D. F., and Bergelson, J. (2003). Negative cross-talk between salicylate- and jasmonate-mediated pathways in the Wassilewskija ecotype of *Arabidopsis thaliana*. *Mol. Ecol.* **12:**1125–1135.

Trebbi, G., Borghini, F., Lazzarato, L., Torrigiani, P., Calzoni, G. L., and Betti, L. (2007). Extremely low frequency weak magnetic fields enhance resistance of NN tobacco plants to tobacco mosaic virus and elicit stress-related biochemical activities. *Bioelectromagnetics* **28:**214–223.

Truman, W., Bennett, M. H., Kubigsteltig, I., Turnbull, C., and Grant, M. (2007). Arabidopsis systemic immunity uses conserved defense signaling pathways and is mediated by jasmonates. *Proc. Natl. Acad. Sci. U. S. A.* **104:**1075–1080.

Ueki, S., and Citovsky, V. (2002). The systemic movement of a tobamovirus is inhibited by a cadmium-ion-induced glycine-rich protein. *Nat. Cell Biol.* **4:** 478–486.

Van der Biezen, E. A., and Jones, J. D. G. (1998). The NB-ARC domain: a novel signalling motif shared by plant resistance gene products and regulators of cell death in animals. *Curr. Biolog.* **8**:R226–R227.

Van Loon, L. C., and Van Kammen, A. (1970). Polyacrylamide disc-electrophoresis of soluble leaf proteins from *Nicotiana tabacum* var. "Samsun" and "Samsun NN". II Changes in protein constitution after infection with tobacco mosaic virus. *Virology* **40:**199–211.

Van Loon, L. C., Pierpoint, W. S., Boller, T., and Conejero, V. (1994). Recommendations for naming plant pathogenesis-related proteins. *Plant Mol. Biol. Reporter* **12:**245–264.

Verberne, M. C., Hoekstra, -J, Bol, J. F., and Linthorst, H. J. M. (2003). Signaling of systemic acquired resistance in tobacco depends on ethylene perception. *Plant J.* **35:**27–32.

Vernooij, B., Friedrich, L., Morse, A., Reist, R., Kolditz-Jawhar, R., Ward, E., Uknes, S., Kessmann, H., and Ryals, J. (1994). Salicylic acid is not the translocated signal responsible for inducing systemic acquired resistance but is required in signal transduction. *Plant Cell* **6:**959–965.

Verma, H. N., and Awasthi, L. P. (1980). Occurrence of a highly antiviral agent in plants treated with *Boerhaavia diffusa* inhibitor. *Can. J. Bot.* **58:**2141–2144.

Verma, H. N., Awasthi, L. P., and Mukerjee, K. (1979). Induction of systemic resistance by antiviral plant extracts in non-hypersensitive hosts. *Z. Pflanzenkrank. Pflanzenschutz* **86:**735–746.

Verma, H. N., Baranwal, V. K., and Srivastava, S. (1998). Antiviral substances of plant origin. *In* "Plant Disease Control" (A. Hadidi, R. K. Khetarpal and H. Koganezawa, eds.), pp. 154–162. The Am. Phytopath. Soc., St. Paul, Minn. U. S. A..

Vidal, S., Cabrera, H., Andersson, R. A., Fredriksson, A., and Valkonen, J. P. (2002). Potato gene Y-1 is an *N* gene homolog that confers cell death upon infection with potato virus Y. *Mol. Plant-Microbe Interact.* **15:**717–727.

Vivek, P., Chowdhury, P. G., and Shalini, S. (2001). Purification of two basic 1,3-beta -glucanase isoforms from *Cyamopsis tetragonoloba* (L.) Taub. induced to resist virus infections. *Israel J. Plant Sci.* **49:**15–19.

Vlot, A. C., Klessig, D. F., and Park, S. W. (2008). Systemic acquired resistance: the elusive signal(s). *Cur. Opin. Plant Biol.* **11:**436–442.

Waller, F., Muller, A., KwiMi, Chung, YunKiam, Yap, Nakamura, K., Weiler, E., and Sano, H. (2006). Expression of a WIPK-activated transcription factor results in increase of endogenous salicylic acid and pathogen resistance in tobacco plants. *Plant Cell Physiol.* **47:**1169–1174.

Wang, B. J., Zhang, Z. G., Li, X. G., Wang, Y. J., He, C. Y., Zhang, J. S., and Chen, S. Y. (2003). Cloning and analysis of a disease resistance gene homolog from soybean. *Acta Bot. Sin.* **45:**864–870.

Waigmann, E, Chen, M.H., Bachmaier, R., Ghoshroy S., and Citovsky V. (2000). Regulation of plasmodesmal transport by phosphorylation of tobacco mosaic virus cell-to-cell movement protein. *EMBO J.* **19**:4875–4884.

Ward, E. R., Uknes, S.. J., Williams, S. C., Dincher, S. S., Wiederhold, D. L., Alexander, D. C., Ahl-Goy, P., Métraux, J. P, and Ryals, J. A. (1991). Coordinate gene activity in response to agents that induce systemic acquired resistance. *Plant Cell* **3:**1085–1094.

Weintraub, M., and Ragetli, H. W. J. (1964). An electron microscope study of tobacco mosaic virus lesions in *Nicotiana glutinosa* L. *J. Cell Biol.* **23:**499–509.

Weststeijn, E. A. (1978). Permeability changes in the hypersensitive reaction of *Nicotiana tabacum* cv. Xanthi n.c. after infection with tobacco mosaic virus. *Physiol. Plant Pathol.* **13:**253–258.

White, R. F., Antoniw, J. F., Carr, J. P., and Woods, R. D. (1983). The effects of aspirin and polyacrylic-acid on the multiplication and spread of TMV in different cultivars of tobacco with and without the *N*-gene. *Phytopathol. Z* **107:**224–232.

Whitham, S., Dinesh-Kumar, S.. P., Choi, D., Hehl, R., Corr, C., and Baker, B. (1994). The product of the tobacco mosaic virus resistance gene N: Similarity to Toll and the interleukin-1 receptor. *Cell* **78:**1101–1115.

Whitham, S. A., Sheng, Q., Chang, H. S., Cooper, B., Estes, B., Zhu, T., Wang, X., and Hou, Y.-M. (2003). Diverse RNA viruses elicit the expression of common sets of genes in susceptible *Arabidopsis thaliana* plan. *Plant J.* **33:**271–283.

Whitham, S., McCormick, S., and Baker, B. (1996). The N gene of tobacco confers resistance to tobacco mosaic virus in transgenic tomato. *Proc. Natl. Acad. Sci. U. S. A.* **93:**8776–8781.

Wright, K. M., Duncan, G. H., Pradel, K. S., Carr, F., Wood., S., Oparka, K. J., and Santa Cruz, S. (2000). Analysis of the *N* gene hypersensitive response induced by a fluorescently tagged tobacco mosaic virus. *Plant Physiol.* **123:**1375–1385.

Wolter, M., Hollricher, K., Salamini, F., and Schulze-Lefert, P. (1993). The *mlo* resistance alleles to powdery mildew infection in barley trigger a developmentally controlled defence mimic phenotype. *Moi. Gen. Genet.* **239:**122–128.

Won, D. H., Sang, H. L., Min, C. K., Jong, C. K., Woo, S. C., Hyun, J. C., Kyoung, J. L., Chan, Y. P., Hyeong, C. P., Ji, Y. C., and Moo, J. C. (1999). Involvement of specific calmodulin isoforms in salicylic acid-independent activation of plant disease resistance responses. *Proc. Natl. Acad. Sci. U. S. A.* **96:**766–771.

Wong, C. E., Carson, R. A., and Carr, J. P. (2002). Chemically induced virus resistance in Arabidopsis thaliana is independent of pathogenesis-related protein expression and the NPR1 gene. *Mol. Plant-Microbe Interact.* **15**:75–81.

Wu, J. H., and Dimitman, J. E. (1970). Leaf structure and callose formation as determinants of TMV movement in bean leaves as revealed by uv irradiation studies. *Virology* **40**:820–827.

Wu, J., and Zhou, X. (2002). Effects of replacing the movement protein gene of *Tobacco mosaic virus* by that of *Tomato mosaic virus*. *Virus Res.* **87**:61–67.

XiaoRong, T., XuePing, Z., GuiXin, L., and JiaLin, Y. (2003). Two amino acids on 2a polymerase of *Cucumber mosaic virus* co-determine hypersensitive response on legumes. *Sci. China Ser. C - Life Sci.* **46**:40–48.

Xie, Z., Fan, B., Chen, C., and Chen, Z. (2001). An important role in an inducible RNA- dependent RNA polymerase in plant antiviral defense. *Proc. Natl. Acad. Sci. U. S. A.* **98**:6516–6521.

Yang, S. J., Carter, S. A., Cole, A. B., Cheng, N. H., and Nelson, R. S. (2004). A natural variant of a host RNA-dependent RNA polymerase is associated with increased susceptibility to viruses by *Nicotiana benthamiana*. *Proc. Natl. Acad. Sci. U. S. A.* **10**:6297–6302.

Yang, Y., and Klessig., D. F. (1966). Isolation and characterization of a tobacco mosaic virus-inducible *myb* oncogene homolog from tobacco. *Proc. Natl. Acad. Sci. U. S. A.* **93**:14972–14977.

Yasuda, M., Nishioka, M., Nakashita, H., Yamaguchi, I., and Yoshida, S. (2003). Pyrazolecarboxylic acid derivative induces systemic acquired resistance in tobacco. *Biosci., Biotechnol. Biochem.* **67**:2614–2620.

Yarwood, C. E. (1960). Localized acquired resistance to tobacco mosaic virus. *Phytopathology* **50**:741–744.

Yoda, H., Ogawa, M., Yamaguchi, Y., Koizumi, N., Kusano, T, and Sano, H (2002). Identification of early-responsive genes associated with the hypersensitive response to tobacco mosaic virus and characterization of a WRKY-type transcription factor in tobacco plants. *Mol. Genet. Genomics* **267**:154–161.

Yoo, B., Kragler, F., Varkonyi-Gasic, E., Haywood, V., Archer-Evans, S., Lee, Y. M., Lough, T. J., and Lucas, W. J. (2004). A systemic small RNA signaling system in plants. *Plant Cell* **16**:1979–2000.

Yu, D., Fan, B., MacFarlane, S. A., and Chen, Z. (2003). Analysis of the involvement of an inducible *Arabidopsis* RNA-dependent RNA polymerase in antiviral defense. *Mol. Plant-Microbe Interact.* **16**:206–216.

Zehnder, G. W., Murphy, J. F., Sikora, E. J., and Kloepper, J. W. (2001). Application of rhizobacteria for induced resistance. *Eur. J. Plant Pathol.* **107**:39–50.

Zhang, Y., Fan, W., Kinkema, M., Li, X., and Dong, X. (1999). Interaction of NPR1 with basic leucine zipper protein transcription factors that bind sequences required for salicylic acid induction of the PR-1 gene. *Proc. Natl. Acad. Sci. U. S. A.* **96**:6523–6528.

Zhang, S., and Klessig, D. F. (1998). N resistance gene–mediated de novo synthesis and activation of a tobacco MAP kinase by TMV infection. *Proc. Natl. Acad. Sci. U. S. A.* 5:7433–7438.

Zhang, S., Liu, Y., and Klessig, D. F. (2000). Multiple levels of tobacco WIPK activation during the induction of cell death by fungal elicitins. *Plant J.* **23:** 339–347.

Zhang, Y. B., and Gui, J-F. (2004). Identification and expression analysis of two IFN-inducible genes in crucian carp (*Carassius auratus* L.). *Gene* **325:**43–51.

Zhang, Z. G., Wang, Y. C., and Zheng, X. B. (2002). Systemic acquired resistance induced by a new kind of 90 kDa extracellular elicitor protein. *Acta Phytopathol. Sinica* **32:**338–346.

Zhou, S. M., Liu, W. N., Kong, L. A., and Wang, M. (2008). Systemic PCD occurs in TMV-tomato interaction. *Scien. China Series C: Life Scien.* **51:**1009–1019.

CHAPTER 4

Recessive Resistance to Plant Viruses

V. Truniger and **M.A. Aranda**

Contents

Abstract

About half of the 200 known virus resistance genes in plants are recessively inherited, suggesting that this form of resistance is more common for viruses than for other plant pathogens. The use of such genes is therefore a very important tool in breeding programs to control plant diseases caused by pathogenic viruses. Over the last few years, the detailed analysis of many host/virus combinations has substantially advanced basic research on

Centro de Edafología y Biología Aplicada del Segura (CEBAS), Consejo Superior de Investigaciones Científicas (CSIC), Apdo Correos 164, 30100 Espinardo (Murcia), Spain

Advances in Virus Research, Volume 75
ISSN: 0065-3527, DOI: 10.1016/S0065-3527(09)07504-6

recessive resistance mechanisms in crop species. This type of resistance is preferentially expressed in protoplasts and inoculated leaves, influencing virus multiplication at the single-cell level as well as cell-to-cell movement. Importantly, a growing number of recessive resistance genes have been cloned from crop species, and further analysis has shown them all to encode translation initiation factors of the 4E (eIF4E) and 4G (eIF4G) families. However, not all of the loss-of-susceptibility mutants identified in collections of mutagenized hosts correspond to mutations in *eIF4E* and *eIF4G*. This, together with other supporting data, suggests that more extensive characterization of the natural variability of resistance genes may identify new host factors conferring recessive resistance. In this chapter, we discuss the recent work carried out to characterize loss-of-susceptibility and recessive resistance genes in crop and model species. We review actual and probable recessive resistance mechanisms, and bring the chapter to a close by summarizing the current state-of-the-art and offering perspectives on potential future developments.

I. INTRODUCTION

Plants that are susceptible to viruses must provide a permissive environment for the viral infection cycle. Viruses must be able to replicate in infected cells, move from cell to cell through plasmodesmata, colonize the whole plant using the vasculature, and interact with vectors that reinitiate infection cycles in other plants. Since the repertoire of viral gene products is typically very small due to limited genome size, viruses need to recruit plant factors to complete the steps of their infection cycle. The plant and viral gene products must cooperate to confer susceptibility and the absence of appropriate plant factors (or their presence in a form that cannot be recognized by the corresponding viral components) may thus confer resistance. This has been termed *passive resistance* because no activity is required by the plant (Fraser, 1990). In genetic terms, host susceptibility factors are encoded by dominant susceptibility alleles and resistance is thus conferred by recessive resistance alleles (Fraser, 1990). This is the main hypothesis to explain recessive resistance to plant viruses, although alternative possibilities exist such as the absence of factors that counteract resistance responses. The latter has yet to be observed for plant–virus interactions, although it is a common feature of plants interacting with fungal pathogens, for example the *mlo* gene in barley (Buschges *et al.*, 1997). In contrast, the cloning and characterization of recessive virus resistance genes in crop species, and loss-of-susceptibility mutants in experimental and crop species, has provided ample evidence to support the former model. This chapter updates our earlier review (Díaz-Pendón *et al.*, 2004) by covering recent work on loss-of-susceptibility and recessive resistance genes in crop and model species, as well as the mechanisms underlying resistance, and future perspectives.

II. LOSS-OF-SUSCEPTIBILITY IN COLLECTIONS OF MUTAGENIZED HOSTS REFLECTS THE MODIFICATION OF TRANSLATION INITIATION FACTORS AND OTHER PLANT PROTEINS

In *Arabidopsis thaliana*, the screening of large mutagenized populations has led to the identification of a number of mutants in which virus susceptibility is reduced or eliminated. Many of the corresponding genes have been cloned and their products characterized (Table 1). TOM1 (and its homolog TOM3) and TOM2A are required for efficient *Tobacco mosaic virus* (TMV) replication in *A. thaliana* protoplasts. They are host integral membrane proteins that localize to the tonoplast, and interact with each other and also with viral replication factors to facilitate the formation of the tobamoviral replication complex (Hagiwara *et al.*, 2003; Ishikawa *et al.*, 1991, 1993; Ohshima *et al.*, 1998; Tsujimoto *et al.*, 2003; Yamanaka *et al.*, 2002). The effect of the *tom1, tom3,* and *tom2A* mutations appears specific for tobamoviruses, and the phenotype can be induced by RNA interference against *Tom1* and *Tom3* in *Nicotiana tabacum*, resulting in near complete inhibition of TMV and other tobamoviruses without affecting plant growth or the replication of the cucumovirus *Cucumber mosaic virus* (CMV) (Asano *et al.*, 2005).

Similarly, virus replication in single *A. thaliana* cells is inhibited in the *lsp1* mutant (Table 1). In this case, the mutation affects the multiplication of at least three viruses of the family *Potyviridae* (here generally described as potyviruses): *Turnip mosaic virus* (TuMV), *Tobacco etch virus* (TEV), and *Lettuce mosaic virus* (LMV). However, there is no effect on another potyvirus, *Clover yellow vein virus* (CLYVV), or on the carmovirus *Turnip crinkle virus* (TCV) (Duprat *et al.*, 2002; Lellis *et al.*, 2002; Sato *et al.*, 2005). Map-based cloning of *Lsp1* revealed that it encodes an isoform of the eukaryotic translation initiation factor 4E (eIF(iso)4E) (Lellis *et al.*, 2002). As we will see below, the 4E family of translation initiation factors is a central component of recessive resistance in many plant–virus interactions. The *A. thaliana cum1-1* mutation inactivates eIF4E, and delays the spread of CMV (but not TMV or TCV) in inoculated leaves, thus inhibiting cell-to-cell movement in a virus-specific manner (Yoshii *et al.*, 1998a, 2004). The same mutation also blocks the accumulation of ClYVV, which accumulates to high levels in wild-type *A. thaliana* (Sato *et al.*, 2005).

TABLE 1 Loss-of-susceptibility (cloned genes) in collections of mutagenized hosts

Host	Virus (Genus; Family)	Allele	Phenotype of mutant	Encoded protein and its function in the viral cycle	References
Arabidopsis thaliana	*Tobacco mosaic virus* (TMV) (*Tobamovirus*)	*tom1-1, tom1-2, tom1-3*	Mutations affect amplification of TMV-related RNAs in a single cell. They do not affect TCV or CMV multiplication.	TOM1 is a transmembrane protein localized in the tonoplast. It interacts with the helicase domain of tobamovirus-encoded replication proteins and is an essential constituent of the tobamoviral replication complex.	Ishikawa *et al.* (1991, 1993), Yamanaka *et al.* (2000), Hagiwara *et al.* (2003)
		tom3	Simultaneous mutations in *Tom3* and *Tom1* are associated with complete inhibition of tobamovirus multiplication. They do not affect TCV or CMV multiplication.	*Tom3* is one of the two *Tom1*-like genes in *Arabidopsis*. There appears to be some functional redundancy between both genes, at least regarding tobamovirus multiplication.	Yamanaka *et al.* (2002)

	tom2A	Mutation affects accumulation of TMV-related RNAs in protoplasts in a tobamovirus-specific manner.	TOM2A is a four-pass transmembrane protein with a C-terminal farnesylation signal. It localizes to the tonoplast, interacts with TOM1, and appears to facilitate the formation of the tobamoviral RNA replication complex in conjunction with TOM1.	Ohshima *et al.* (1998), Tsujimoto *et al.* (2003), Hagiwara *et al.* (2003)
Cucumber mosaic virus (CMV) *(Cucumovirus; Bromoviridae)*	*cum1-1*	Mutation affects spreading of CMV within an infected leaf, delaying the CMV cell-to-cell movement in a virus-specific manner (i.e., it does not affect movement of TCV or TMV). Accumulation of CMV RNAs in inoculated mutant	Translation factor eIF4E1. The expression of the CMV movement protein (3a) is reduced in inoculated mutant protoplasts, probably due to a reduced translational efficiency of its messenger RNA (CMV RNA3); this	Yoshii *et al.* (1998a, 2004), Sato *et al.* (2005)

Table 1 *(Continued)*

Host	Virus (Genus; Family)	Allele	Phenotype of mutant	Encoded protein and its function in the viral cycle	References
			protoplasts is not affected.	effect depends of the 5′- and 3′-control regions of CMV RNAs.	
			Multiplication of *Clover yellow vein virus* (ClYVV; *Potyvirus; Potyviridae*) was not detected in this mutant; however, this mutation did not affect multiplication of another potyvirus, TuMV.		
	Cucumber mosaic virus (Cucumovirus; Bromoviridae) Turnip crinkle virus (TCV)	*cum2-1*	Mutation affects spreading of CMV (but not that of TMV) within an infected leaf, delaying the CMV cell-to-cell	Translation factor eIF4G. Its function seems to be indistinguishable from that of eIF4E1 in relation to CMV movement (see	Yoshii *et al.* (1998b, 2004)

(Carmovirus; Tombusviridae)			movement. CMV RNA accumulation in protoplasts is not affected. This mutation also affects spreading of TCV. Contrarily to CMV, accumulation of TCV RNAs and proteins is drastically reduced in inoculated protoplasts.	above). In the case of TCV, the mutation affects translational efficiency of uncapped RNAs carrying the TCV 3′-untranslated region.	
Tobacco etch virus (TEV) *(Potyvirus; Potyviridae)* *Turnip mosaic virus* (TuMV) *(Potyvirus; Potyviridae)*	*lsp1*		Mutants are defective in supporting TuMV and TEV genome expression and/or replication. An equivalent mutant also showed resistance to *Lettuce mosaic virus* but not to *Clover yellow vein virus*; both viruses belong to the genus *Potyvirus*.	Translation factor eIF (iso)4E. It interacts with potyviral VPg. The interaction of eIF(iso)4E with VPg-Pro takes place in the cellular nucleus, whereas the interaction with 6K-VPg-Pro takes place in cytoplasmic vesicles, suggesting that association of eIF(iso)4E with VPg might be needed for two different	Wittmann *et al.* (1997), Whitham *et al.* (1999), Lellis *et al.* (2002), Duprat *et al.* (2002), Sato *et al.* (2005), Beauchemin *et al.* (2007)

Table 1 *(Continued)*

Host	Virus (Genus; Family)	Allele	Phenotype of mutant	Encoded protein and its function in the viral cycle	References
				functions, depending on the VPg precursor involved.	
Oryza sativa	*Rice dwarf virus* (RDV) *(Phytoreovirus; Reoviridae)*	*rim1-1*	Virus accumulation in mutants is drastically reduced, but not suppressed. The mutant is susceptible to rice viruses other than RDV. Non-inoculated mutant plants show a phenotype consisting of slow growth, short roots, and sometimes twisting of leaf tips.	Novel NAC-domain protein. The expression levels of its mRNA do not change significantly after infection. It works as a transcriptional activator, at least in yeast cells. Its overexpression associates with increased virus accumulation, but its role in the virus cycle is not known.	Yoshii *et al.* (2009)

The *A. thaliana cum2-1* mutant inactivates eIF4G. This also delays the spread of CMV in inoculated leaves, but the accumulation of CMV RNAs in inoculated mutant protoplasts is not affected (Yoshii *et al.*, 1998b). Unlike *cum1-1*, the *cum2-1* mutant is also associated with resistance to TCV, and the accumulation of viral RNAs and proteins is blocked in inoculated mutant protoplasts (Yoshii *et al.*, 1998b).

Screens for loss-of-susceptibility mutants have also been carried out in other host plants. A recent example is the identification of the rice mutant *rim1-1*, in which the accumulation of *Rice dwarf virus* (RDV) is drastically reduced, although not completely suppressed. This mutant remains susceptible to other rice viruses, and the resistance phenotype reflects a mutation in a gene encoding a novel NAC-domain protein, although the role of this protein in the virus infection cycle is not yet known (Yoshii *et al.*, 2009). Interestingly, at least two phylogenetically diverse plant RNA viruses, *Brome mosaic virus* (BMV) and *Tomato bushy stunt virus* (TBSV), are able to replicate in the yeast *Saccharomyces cerevisiae* (Janda and Ahlquist, 1993; Panavas and Nagy, 2005) allowing powerful yeast resources to be used in genome-wide mutant screens (Kushner *et al.*, 2003; Nagy and Pogany, 2006). Numerous loss-of-susceptibility yeast mutants have been identified and characterized (Nagy, 2008). Although artificially induced mutations in yeast may help to increase our understanding of virus infection mechanisms, and may eventually be transferable to crop plants, such data are outwith the scope of this chapter and we refer the reader to the recent review by Nagy for further information (Nagy, 2008).

III. ALL KNOWN RECESSIVE RESISTANCE GENES IN CROP SPECIES ENCODE TRANSLATION INITIATION FACTORS

Kang and co-workers (2005b) have provided a list of more than 200 published virus resistance genes. About half of them are recessively inherited, which suggests that recessive resistance is more common for plant viruses than for other plant pathogens, where resistance appears to be predominantly inherited as a monogenic dominant trait (Fraser, 1990). Moreover, recessive resistance appears to be more frequent for potyviruses than for viruses of other families, as more than half of the recessive resistance genes listed by Kang and co-workers (2005b) provide resistance to viruses of this family. It is possible that more research has been carried out on resistance to potyviruses than other virus families, but there may also be genuine peculiarities of potyvirus biology that make recessive resistance to these viruses more likely. We will return to this subject later in the chapter.

Since our last review (Díaz-Pendón *et al.*, 2004), basic research on the control of and mechanisms underlying recessive resistances in crop species has advanced significantly. Resistance has been studied in detail for more than 25% of the host/virus combinations listed in recent reviews (Díaz-Pendón *et al.*, 2004; Kang *et al.*, 2005b). In most cases, resistance was expressed in protoplasts, in inoculated leaves, and/or appeared to affect the cell-to-cell movement of the virus (Díaz-Pendón *et al.*, 2004; Kang *et al.*, 2005b) (Table 2). Importantly, a large number of recessive resistance genes have been cloned from crop species, and all of them encode translation initiation factors of the 4E or 4G families (Robaglia and Caranta, 2006) (Table 2). The first was *pvr2*, which confers resistance to *Potato virus Y* (PVY) in pepper (Ruffel *et al.*, 2002). This was followed by *mo1*, which confers resistance to LMV in lettuce (Nicaise *et al.*, 2003). In both cases, a candidate gene approach identified *eIF4E* as the resistance gene. The candidate was chosen based on the previous results obtained with *A. thaliana lsp1*, which confers resistance to TEV/TuMV/LMV (see above), and also based on a number of results related to the properties of the viral genome-linked protein VPg, which was shown to act as a potyviral avirulence factor for several host/potyvirus combinations (Table 2). Other recessive genes conferring resistance to potyviruses in crop species encode eIF4E and its isoform eIF(iso)4E (Table 2). Perhaps more surprisingly, this translation initiation factor has also been shown to control recessive resistance to the carmovirus *Melon necrotic spot virus* (MNSV) (Nieto *et al.*, 2006). Furthermore, recessive resistance to the sobemovirus *Rice yellow mottle virus* (RYMV) was shown to be controlled by eIF(iso)4G (Albar *et al.*, 2006).

The roles of these factors in the uninfected plant cell provide important insights into their role in the virus infection cycle. In plant cells, there are two cap-binding complexes, eIF4F and eIF(iso)4F. The first is formed by eIF4E, a small cap-binding protein that binds the m^7G group (the cap structure) on the 5′-ends of mRNAs, and eIF4G, a large protein that interacts with other initiation factors, including the poly(A) binding protein (PABP), eIF4A, and the multisubunit factor eIF3 (Browning, 2004). The interaction between eIF4F and mRNA is thought to be the first step in the initiation of translation, where eIF4E provides the 5′-cap-binding function during the formation of translation initiation complexes on most eukaryotic mRNAs. In addition, circularization of the mRNA is necessary for efficient translation (Kawaguchi and Bailey-Serres, 2002), and this requires a triple interaction in which eIF4G acts as a scaffold, interacting with eIF4E (which binds the cap structure) and PABP (which binds the poly(A) tail). Like eIF4F, eIF(iso)4F also comprises two subunits, the small cap-binding protein eIF(iso)4E, and the large eIF(iso)4G subunit. Both eIF4F and eIF(iso)4F have similar *in vitro* activities (Browning, 1996) and there is evidence that both are

TABLE 2 Recessive resistances (cloned genes) in crop species

Host	Sources of resistance	Virus (Genus; Family)	Allele	Coding for	Expression	Viral counterpart	Selected references
Capsicum annuum	Double haploid line (DH801) recovered from the F-1 hybrid between two *C. annuum* lines, Perennial and Florida VR2.	*Pepper veinal mottle virus* (PVMV) *(Potyvirus; Potyviridae)*	$pvr2^2$+ *pvr6*	$pvr2^2$: eIF4E *pvr6*: eIF(iso)4E Separately, they do not confer resistance.	Undetectable virus accumulation in inoculated leaves.		Caranta *et al.* (1996), Ruffel *et al.* (2006)
	Cultivars Yolo Y, Avelar, Florida VR2 and other accessions.	*Potato virus Y* (PVY) *(Potyvirus; Potyviridae)*	$pvr2^1$, $pvr2^2$ *and* $pvr2^3$ *to* $pvr2^9$	eIF4E	Undetectable virus accumulation in inoculated leaves due to impaired cell-to-cell movement of the virus.	Mutations in VPg abolish resistance. VPg interacts with eIF4E and this interaction correlates with susceptibility.	Dogimont *et al.* (1996), Arroyo *et al.* (1996), Ruffel *et al.* (2002), Moury *et al.* (2004), Charron *et al.* (2008)
	Dempsey, Florida VR2 and Chay Angolano.	*Tobacco etch virus* (TEV) *(Potyvirus; Potyviridae)*	$pvr2^2$	eIF4E	Undetectable virus accumulation in inoculated leaves or protoplasts.	Interaction of VPg variants with eIF4E in the yeast two-hybrid system correlates with susceptibility.	Deom *et al.* (1997), Kang *et al.* (2007), Charron *et al.* (2008)

Table 2 *(Continued)*

Host	Sources of resistance	Virus (Genus; Family)	Allele	Coding for	Expression	Viral counterpart	Selected references
Capsicum chinense	PI 152225, PI 159236	*Pepper mottle virus* (PepMoV) *(Potyvirus; Potyviridae)*	*pvr1*	eIF4E	Undetectable virus accumulation in inoculated leaves or protoplasts.		Murphy and Kyle (1995), Deom *et al.* (1997), Kyle and Palloix (1997), Murphy *et al.* (1998)
	PI 159236	*Potato virus Y* (PVY) *(Potyvirus; Potyviridae)*	*pvr1*	eIF4E	Undetectable virus accumulation in inoculated leaves.		Boiteux *et al.* (1996)
	PI 152225, PI 159236	*Tobacco etch virus* (TEV) *(Potyvirus; Potyviridae)*	*pvr1*	eIF4E	Undetectable virus accumulation in inoculated leaves or protoplasts.	Mutations in VPg abolish resistance. VPg interacts with eIF4E and this interaction correlates with susceptibility.	Murphy and Kyle (1995), Deom *et al.* (1997), Murphy *et al.* (1998), Kang *et al.* (2005a, 2007), Yeam *et al.* (2007)
Cucumis melo	Gulfstream, PI 161375, Planters Jumbo and other accessions	*Melon necrotic spot virus* (MNSV) *(Carmovirus; Tombusviridae)*	*nsv*	eIF4E	Undetectable virus accumulation in inoculated leaves or protoplasts. Temperature-sensitive inactivation occurs when plants/protoplasts are grown below 20 °C.	A non-coding region at the 3′-end of the viral genome, which contains a cap-independent translational enhancer, controls virulence.	Diaz *et al.* (2004), Nieto *et al.* (2006, 2007), Kido *et al.* (2008), Truniger *et al.* (2008)

Hordeum vulgare	Franka, Ym no. 1, Miho Golden and other accessions	*Barley mild mosaic virus* (BaMMV) *(Bymovirus; Potyviridae)*	*rym 4*, *rym 5* (and *rym* 6)	eIF4E	Immunity.	Resistance-breaking capacity attributed to mutations in the VPg gene.	Kanyuka *et al.* (2004, 2005), Stein *et al.* (2005)
	Express, Carola, Igri, Franka, Mokusekko 3, Hsingwuke 2, and other accessions.	*Barley yellow mosaic virus* (BaYMV) *(Bymovirus; Potyviridae)*	*rym 4*, *rym 5*	eIF4E	Immunity.	Resistance-breaking capacity correlate with mutations in the VPg gene.	Kuhne *et al.* (2003), Kanyuka *et al.* (2005), Stein *et al.* (2005)
Lactuca sativa	Gallega de invierno, Floribibb, Mantilia, Malika, Egyptian wild lines, Salinas 88, Vanguard 75 and other accessions.	*Lettuce mosaic virus* (LMV) *(Potyvirus; Potyviridae)*	$mo1^1$, $mo1^2$	eIF4E	Depends on the combination of resistance allele and genetic background: $mo1^1$ is normally responsible of reduced accumulation and $mo1^2$of symptom alleviation.	Mutations in VPg allow $mo1^1$ to be overcome, mutations in the C-terminal portion of the cylindrical inclusion protein allowed $mo1^1$ and $mo1^2$ to be overcome. Central domain of VPg interacts with eIF4E and Hc-Pro. The ability of eIF4E to bind a cap analogue or to fully interact with eIF4G appeared unlinked to LMV infection.	Nicaise *et al.* (2003), Roudet-Tavert *et al.* (2007), German-Retana *et al.* (2008), Abdul-Razzak *et al.* (2009)

Table 2 *(Continued)*

Host	Sources of resistance	Virus (Genus; Family)	Allele	Coding for	Expression	Viral counterpart	Selected references
Oryza glaberrima and *Oryza sativa*	Gigante, Bekarosaka (*O. sativa*), and Tog 5681 (*O. glaberrima*)	*Rice yellow mottle virus* (RYMV) (*Sobemovirus*)	*rymv-1*	eIF(iso)4G	Impaired movement. Accumulation in protoplasts is indistinguishable in susceptible and resistant varieties.	Resistance-breaking capacity correlate with mutations in the VPg gene.	Albar *et al*. (2003, 2006), Hébrard *et al*. (2006, 2008)
Pisum sativum	JI1405, PI193835 and PI269818	*Pea seed borne mosaic virus* (PSbMV) (*Potyvirus; Potyviridae*)	*sbm1* (and *sbm1*^{1}),	eIF4E	eIF4E complements virus multiplication in primary target cells and virus movement and expression in nearby cells.	Resistance-breaking capacity correlate with mutations in the VPg gene.	Johansen *et al*. (2001), Gao *et al*. (2004), Keller *et al*. (1998)
	PI193835 and PI347464	*Bean yellow mosaic virus* (BYMV) (*Potyvirus; Potyviridae*)	*wlv (=sbm1)*	eIF4E	Systemic resistance.	Resistance-breaking capacity correlate with mutations in the VPg gene.	Provvidenti and Hampton (1993), Bruun-Rasmussen *et al*. (2007)
Solanum hirsutum	PI247087	*Potato virus Y* (PVY) (*Potyvirus; Potyviridae*)	*pot-1*	eIF4E	Undetectable virus accumulation in inoculated leaves.	Resistance-breaking capacity correlate with mutations in the VPg gene.	Parrella *et al*. (2002), Moury *et al*. (2004), Ruffel *et al*. (2005)

PI247087	*Tobacco etch virus* (TEV) *(Potyvirus; Potyviridae)*	*pot-1*	eIF4E	Undetectable virus accumulation in inoculated leaves.	Strain-specific interaction of VPg variants with tomato eIF4E in the yeast two-hybrid system.	Schaad *et al.* (2000), Parrella *et al.* (2002), Ruffel *et al.* (2005)

functionally interchangeable, but capable of discriminating between different mRNAs *in vitro*. Whereas eIF(iso)4F promotes translation preferentially from unstructured mRNAs, eIF4F can promote translation from uncapped and structured mRNAs alike (Gallie and Browning, 2001). There must be a degree of functional redundancy for eIF4E functions, since mutant *A. thaliana* and pepper plants lacking one of the two isoforms show no obvious phenotype (Duprat *et al.*, 2002; Ruffel *et al.*, 2006; Sato *et al.*, 2005; Yoshii *et al.*, 2004), even though the factors are differentially expressed in *A. thaliana* tissues (Leonard *et al.*, 2004). In addition to what seems to be its fundamental role in translation initiation, eIF4E has been shown to accumulate within the nucleus in nuclear bodies, where it is involved in the export of a subset of mRNAs containing a structure known as a 4E-sensitivity element (reviewed by Goodfellow and Roberts, 2008).

IV. POTYVIRUS RESISTANCE MEDIATED BY EIF4E/EIF(ISO)4E

Several mutations in eIF4E and eIF(iso)4E provide resistance to potyvirus infection in a range of hosts (Robaglia and Caranta, 2006) (see Tables 1 and 2). In nearly all cases, resistance results in undetectable virus multiplication in inoculated leaves, as shown for TuMV, *Plum pox virus* (PPV) and LMV (*lsp1*), *Barley yellow mosaic virus* (BaYMV) and *Barley mild mosaic virus* (BaMMV) (*rym4-6*), PVY (*pvr1, pot1*), TEV (*pvr2*^{2}*/1, pot1*), *Pepper veinal mottle virus* (PVMV) (*pvr2* and *pvr6*), and *Pea seed-borne mosaic virus* (PSbMV) and *Bean yellow mosaic virus* (BYMV) (*sbm1*) (Bruun-Rasmussen *et al.*, 2007; Decroocq *et al.*, 2006; Duprat *et al.*, 2002; Gao *et al.*, 2004; Kang *et al.*, 2005a; Kanyuka *et al.*, 2004; Kuhne *et al.*, 2003; Ruffel *et al.*, 2005, 2006). This form of resistance can be caused by a defect in virus multiplication in single cells, in cell-to-cell movement, or both. To distinguish these possibilities, virus multiplication has been analyzed in inoculated protoplasts and also by monitoring the progress of infections using GFP- or GUS-expressing infectious clones. For example, in protoplast experiments, the resistance of pepper (*pvr2*^{2}) to TEV has been shown to act at the single-cell level (Deom *et al.*, 1997), whereas the resistance conferred by *pvr2*^{1} is associated with a defect in PVY cell-to-cell movement (Arroyo *et al.*, 1996; Kang *et al.*, 2005a). Using reporter gene-expressing infectious clones, no single-cell infection could be detected for LMV (*mo1*), TEV (*lsp1*), ClYVV (*cum1-1*), or PSbMV (*sbm1*^{1}) (Candresse *et al.*, 2002; Gao *et al.*, 2004; Lellis *et al.*, 2002; Sato *et al.*, 2003, 2005). However, it is not always possible to make a clear-cut distinction between resistance affecting single-cell accumulation and cell-to-cell virus movement. For example, using a PSbMV-P1 isolate expressing GFP, the pea *sbm1* resistance gene was shown to affect virus multiplication at the single-cell level. In agreement with this, the susceptibility allele not only *trans*-complemented single-cell virus multiplication in resistant plants, but also allowed virus trafficking from one cell to another (Gao *et al.*, 2004). In a few cases, resistance is manifested as reduced virus accumulation in whole plants (e.g., lettuce *mo1*^{2} resistance to LMV) (Candresse *et al.*, 2002; German-Retana *et al.*, 2000; Revers *et al.*, 1997). Therefore, most examples of eIF4E/eIF(iso)4E-mediated resistance seem to affect the early steps of potyvirus infection, although the actual step or steps affected are not necessarily the same.

Different potyviruses appear to exploit different members of the eIF4E family for multiplication (Tables 1 and 2). In *A. thaliana*, for example, the loss of eIF4E confers resistance to ClYVV (Sato *et al.*, 2005), but the loss of eIF(iso)4E confers resistance to TuMV, TEV, and LMV (Duprat *et al.*, 2002; Lellis *et al.*, 2002). Interestingly, some potyviruses use different eIF4E isoforms for multiplication in different hosts. For example, TEV and LMV need eIF(iso)4E for successful infection of *A. thaliana* (Duprat *et al.*, 2002; Lellis *et al.*, 2002), whereas the same viruses need eIF4E in pepper and in tomato, and lettuce, respectively (Kang *et al.*, 2005a; Nicaise *et al.*, 2003; Ruffel *et al.*, 2005). Although most potyviruses need one specific eIF4E isoform to multiply in a specific host, others can use both of them, as appears to be the case for PVMV in pepper (Ruffel *et al.*, 2006). Furthermore, different alleles can cause different levels of resistance in a given host. For example, while no single-cell infection foci were detected in the leaves of *A. thaliana lsp1-1* mutants inoculated with TuMV expressing GFP or TEV expressing GUS, single-cell infection foci were detected in *lsp1-2* mutant leaves inoculated with the TEV construction, albeit rarely. These two alleles differ in having stop codons at positions 63 (*lsp1-1*) and 120 (*lsp1-2*) (Lellis *et al.*, 2002). Interestingly, some resistance alleles provide different levels of resistance depending on the virus or even its pathotype. The pepper *pvr2*^{1} resistance gene, for example, was shown to affect PVY cell-to-cell movement but block TEV replication (Arroyo *et al.*, 1996; Deom *et al.*, 1997; Kang *et al.*, 2005a). Also, the pea *sbm1*^{1} resistance gene confers resistance to PSbMV P1 in protoplasts, but not to PSbMV P4 (Keller *et al.*, 1998). Similarly, for LMV/lettuce *mo1*, resistance depends on the viral isolate and host genetic background (Revers *et al.*, 1997). In conclusion, potyviruses demonstrate notable isoform specificity, with eIF4E and eIF(iso)4E usually having distinct and non-overlapping functions in their infection cycle. The interaction between potyviruses and eIF4E/eIF(iso)4E appears to control both qualitative and quantitative aspects of viral multiplication.

A. Mutations in eIF4E that are responsible for potyvirus resistance

The diverse phenotypes associated with eIF4E-mediated potyvirus resistance result from one or a small number of amino acid changes in the eIF4E sequence (Fig. 1a). Single amino acid substitutions include mutation Ala70Pro in the *mo1*^{2} allele from lettuce (Nicaise *et al.*, 2003), Val67Glu in the *pvr2*^{4} allele from pepper (Charron *et al.*, 2008), and Gly107Arg in *pvr1* also from pepper (Yeam *et al.*, 2007). Very similar substitutions at corresponding positions exist in the proteins encoded by the *sbm1* allele in pea (Gly107Arg) and the *mo1*^{1} allele in lettuce (Gly109His), but these alleles also contain additional mutations. Even within a single plant, different *eIF4E* alleles exist that cause resistance to the same and/or different virus isolates (Charron *et al.*, 2008; Nicaise *et al.*, 2003).

The crystal structure has been solved for several different eIF4Es, including those from wheat and pea (Ashby *et al.*, personal communication; Monzingo *et al.*, 2007), pinpointing the location of crucial amino acid substitutions responsible for resistance in the tertiary structure of the protein. The wheat eIF4E structure has much in common with the mammalian and yeast forms of eIF4E, with eight β-strands, three α-helices, and three extended loops. Wild-type plant eIF4E proteins contain an intramolecular disulfide bridge, which is not present in other eukaryotes. Only the wheat eIF4E mutant Cys113Ser, unable to form this disulfide bridge, could be crystallized with m^7-GDP in its cap-binding pocket (Fig. 1b) (the wild-type protein could not be co-crystallized with m^7-GDP). As can be seen in Fig. 1b, in the crystal structure of wheat eIF4E (Monzingo *et al.*, 2007) mutations causing eIF4E-mediated resistance cluster near the cap-recognition pocket, some inside and others on the outer surface of the fingers forming the pocket, especially at the right-hand side (Kang *et al.*, 2005a; Nicaise *et al.*, 2003; Ruffel *et al.*, 2005; Stein *et al.*, 2005). Only two residues that are mutated in resistance alleles are also involved in cap binding (colored pink in Fig. 1b). The highly conserved amino acid residues thought to be involved in stabilizing the protein structure (Monzingo *et al.*, 2007) are not mutated in eIF4E alleles that confer resistance (not shown). Interestingly, the mutation of Gly107 in pepper eIF4E (colored light blue in Fig. 1b) affects the protein's cap-binding activity and this correlates with its location in the cap-binding pocket, while the mutation of Val67 in the same protein (blue in Fig. 1b) does not affect cap binding, reflecting its location on the outer surface (Yeam *et al.*, 2007).

In contrast to *eIF4E* mutations conferring resistance, which cause amino acid substitutions, the only resistance-conferring eIF(iso)4E alleles described thus far correspond to knock-outs (*lsp1*) or truncated proteins

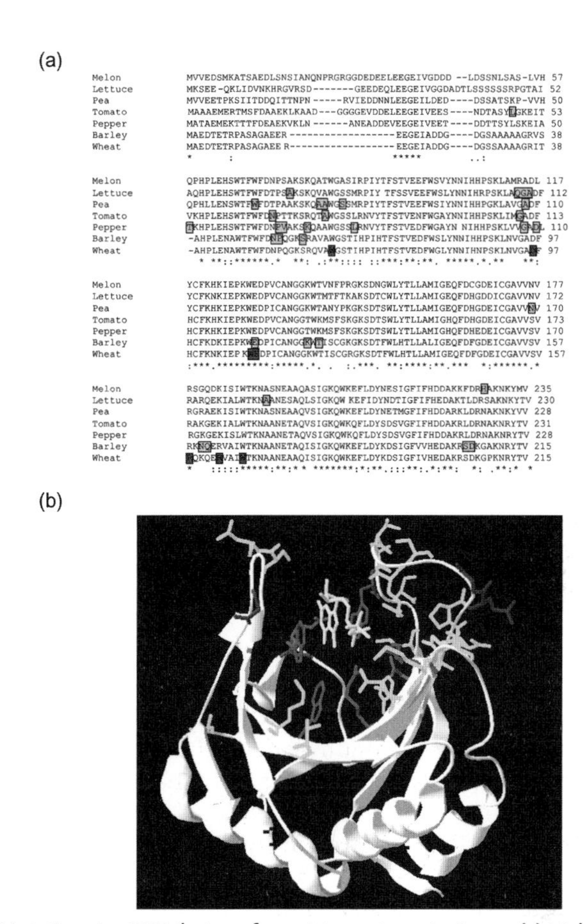

FIGURE 1 Mutations in eIF4E that confer resistance to potyviruses. (a) Multiple sequence alignment of eIF4E from different plant species. Amino acids highlighted in green show mutations that confer virus resistance in lettuce (AAP86602), pea (AAR04332), tomato (AAF70507), pepper (AAS68034), and barley (AAV80393). Additionally, the unique mutation that confers resistance to MNSV in melon is shown in green. Amino acids highlighted in red in the wheat eIF4E sequence (CAA78262) are those identified in the crystal structure to be most directly involved in binding m^7-GTP (Monzingo *et al.*, 2007). (b) Positions of amino acids corresponding to residues involved in eIF4E-mediated resistance from different plant species in the three-dimensional structure of wheat Cys113Ser eIF4E (NCBI structure 2IDV) (Monzingo *et al.*, 2007). Affected amino acids are generally highlighted in green, with the exception of residues Asp96 and Glu109 which are highlighted in pink to show they are not only mutated in pepper ($pvr2^2$) and barley eIF4E (rym6) proteins, respectively, but also involved in cap binding (other cap-binding residues colored in red). Amino acids corresponding to single amino acid substitutions causing eIF4E-mediated resistance are shown in light blue (Gly94 corresponding to Gly107 in pepper and pea, and also corresponding to Gly109 in lettuce) and blue (Gln54 and Gly55 corresponding to Val67 in pepper and to Ala70 in lettuce, respectively). Gly208, the only amino acid changed in melon *nsv*, is shown in purple. The m^7-GDP crystallized in the cap-binding pocket is colored yellow. (See Page 4 in Color Section at the back of the book)

(*pvr6*). These deletion mutants seem to confer a more general antiviral resistance (Duprat *et al.*, 2002; Lellis *et al.*, 2002).

B. Resistance-breaking potyvirus isolates

Resistance-breaking isolates have been characterized for several eIF4E/eIF(iso)4E-mediated potyvirus resistance traits (see Tables 1 and 2). The viral determinants required to overcome resistance have, with one recently published exception, always been identified as the virus-encoded protein covalently linked to the 5′ end of the potyvirus genome (VPg), as shown for TuMV (Charron *et al.*, 2008), PVY (Moury *et al.*, 2004), PSbMV (Borgstrom and Johansen, 2001), BYMV (Bruun-Rasmussen *et al.*, 2007), BaYMV (Kuhne *et al.*, 2003), BaMMV (Kanyuka *et al.*, 2004), and LMV (Abdul-Razzak *et al.*, 2009). Very recently, a single amino acid substitution in the C-terminal region of the LMV CI protein has been shown to overcome the lettuce resistance alleles $mo1^1$ and $mo1^2$ (Abdul-Razzak *et al.*, 2009). This is the sole example of eIF4E/eIF(iso)4E-mediated potyvirus resistance being overcome with an avirulence determinant other than VPg.

VPg is required for potyvirus infectivity (Murphy *et al.*, 1996), although its precise role is still unclear. It is produced, like other potyvirus proteins, by the proteolytic cleavage of a polyprotein translated from the virus genomic RNA (Revers *et al.*, 1999). This processing gives rise to different VPg-containing polypeptides, including the precursors VPg-Pro and 6K-VPg-Pro, which have been shown to exist *in planta* (Beauchemin *et al.*, 2007; Leonard *et al.*, 2004; Thivierge *et al.*, 2008). VPg seems to interact with multiple virus and host products. A role in virus RNA synthesis has been suggested, since VPg interacts with the virus RNA polymerase (Fellers *et al.*, 1998; Hong *et al.*, 1995) and has been proposed to act as a replication primer analogous to the mechanism used in picornaviruses (Puustinen and Makinen, 2004; Sharma *et al.*, 2005). PVY VPg has been shown to control the accumulation of virus particles and phloem loading in potato (Rajamäki and Valkonen, 2002) and to translocate to the host cell nucleus (Carrington *et al.*, 1991). *In planta*, 6K-VPg-Pro is targeted to cytoplasmic vesicles embedded in the ER, where it interacts with viral and cellular factors including eIF4E, whereas VPg-Pro is targeted to the nucleus, where it again interacts with eIF4E and other cellular factors (Beauchemin and Laliberte, 2007; Beauchemin *et al.*, 2007; Leonard *et al.*, 2004; Thivierge *et al.*, 2008). The VPgs of resistance-breaking isolates all have one or several mutations in the central region of the protein, which is thought to be exposed on the surface (Roudet-Tavert *et al.*, 2007), indicating that this region could be involved in interacting with eIF4E/eIF(iso)4E. Importantly, the amino acids in the central region of PVY and PVA VPg have undergone positive selection, suggesting that this region's effect on virulence is determined by the protein and not by the virus RNA (Moury *et al.*, 2004). The structures of PVY and PVA VPg are highly disordered, which may explain the multiple functions of this protein in the potyvirus life cycle (Grzela *et al.*, 2008; Rantalainen *et al.*, 2008).

The potyvirus CI protein also plays an important role in infection. It has been shown to be required for replication and systemic movement (Fernandez *et al.*, 1997; Gómez de Cedrón *et al.*, 2006), since specific C-terminal mutations abolished these functions (Carrington *et al.*, 1998; Kekarainen *et al.*, 2002).

C. Interaction between eIF4E/eIF(iso)4E and VPg

Whatever the mechanisms underlying the different eIF4E/eIF(iso)4E-mediated resistance phenotypes, it is likely that some of the functions of VPg in the virus infection cycle depend on its interaction with host factors, amongst them eIF4E/eIF(iso)4E. Using the yeast two-hybrid system and enzyme-linked immunosorbent assays (ELISAs), TuMV VPg and *A. thaliana* eIF(iso)4E have been shown to interact (Wittmann *et al.*, 1997) and this interaction was shown to correlate with virus infectivity (Leonard *et al.*, 2000). Similarly, the interaction between TEV VPg and the eIF4E from susceptible tomato varieties was demonstrated using the yeast two-hybrid system (Schaad *et al.*, 2000), while recombinant proteins and ELISAs were used to show that LMV VPg interacts with lettuce eIF4E (Roudet-Tavert *et al.*, 2007). Yeast two-hybrid and GST pull-down assays showed that the interaction between TEV VPg and eIF4E from susceptible pepper varieties was strong, while eIF4E from resistant varieties (pvr1, $pvr2^1$, and $pvr2^2$) failed to bind TEV VPg (Kang *et al.*, 2005a). Taken together, such data make it clear that mutations in eIF4E/eIF(iso)4E conferring resistance do so by disrupting its interaction with VPg. Interestingly, the amino acid substitutions, Gly107Arg (*pvr1*) and Val67Glu ($pvr2^4$), in pepper eIF4E have independently been shown to be sufficient to prevent interactions with the VPg from TEV (Yeam *et al.*, 2007) and PVY (Charron *et al.*, 2008), respectively. This interaction was shown to occur *in planta* with 6K-VPg-Pro, a precursor form of TuMV VPg that binds eIF4E and eIF(iso)4E in cytoplasmic vesicles embedded in the ER (Beauchemin *et al.*, 2007; Leonard *et al.*, 2004). Since different amino acids are substituted to generate resistant forms of eIF4E in different and even the same plant species, it is thought that the precise contact point between VPg and eIF4E needs to be optimized for each potyvirus (Charron *et al.*, 2008). In the VPgs of all resistance-breaking isolates of different viruses, mutations have been localized to the central region, which is likely to be involved in eIF4E binding (Charron *et al.*, 2008; Roudet-Tavert *et al.*, 2007). An interaction between PSbMV VPg and eIF4E from pea could not be detected using the yeast two-hybrid system (Gao *et al.*, 2004), although this VPg is the avirulence determinant for overcoming eIF4E-mediated resistance conferred by the *sbm1* allele (Borgstrom and Johansen, 2001; Keller *et al.*, 1998). This may reflect the possibility that other host or virus products are required for the interaction.

Mutations in eIF4E/eIF(iso)4E that confer resistance do so by abolishing interactions with VPg, thus preventing infection. However, viruses can evolve to overcome resistance by undergoing compensatory mutations in VPg that restore such interactions. For example, the VPg of resistance-breaking PVY isolates can interact with the product of the *pvr2* allele (Moury *et al.*, 2004). This is supported by the coevolution found to exist between pepper eIF4E and potyvirus VPgs (Charron *et al.*, 2008). Although resistance-breaking mutations in VPg usually restore interactions with the same eIF4E that originally mutated to confer resistance, resistance-breaking is occasionally achieved through the formation of *de novo* interactions with another isoform. For example, a single amino acid change in TuMV VPg is sufficient to overcome the *A. thaliana* eIF(iso)4E null allele (Charron *et al.*, 2008).

D. Cap-binding ability of eIF4E/eIF(iso)4E mutants

There is no consensus on whether or not the VPg- and cap-binding domains of eIF4E overlap. The pepper pvr1 protein can bind neither VPg nor a cap analog, suggesting that the binding sites might overlap (Khan *et al.*, 2006; Miyoshi *et al.*, 2006; Yeam *et al.*, 2007), but the pvr2 protein binds the cap normally despite its inability to bind VPg, suggesting that the overlap is not complete (Yeam *et al.*, 2007). Lettuce eIF4E has similar binding affinities for a cap analog and VPg, but binding was shown to occur at different sites (Michon *et al.*, 2006), whereas *A. thaliana* eIF(iso)4E binds TuMV VPg with higher affinity than capped RNA (Miyoshi *et al.*, 2006). In several cases, cap analogs and VPg compete to bind eIF4E (Khan *et al.*, 2006; Khraiwesh *et al.*, 2008; Leonard *et al.*, 2000). Therefore, the cap-binding ability of eIF4E/eIF(iso)4E does not necessarily correlate with its VPg-binding ability, since some mutant eIF4E/eIF(iso)4E proteins that confer resistance due to their limited interaction with VPg retain their cap-binding ability, while others do not (German-Retana *et al.*, 2008). Because the precise contact point between VPg and eIF4E is optimized for each potyvirus, even in the same plant species (Charron *et al.*, 2008), the various mutations in eIF4E/eIF(iso)4E may have different effects on cap binding.

E. Molecular mechanism(s) of eIF4E/eIF(iso)4E-mediated potyvirus resistance

Several models have been put forward to explain the mechanisms of eIF4E/eIF(iso)4E-mediated potyvirus resistance, including the inhibition of translational initiation or replication, the coupling of both processes, and/or cell-to-cell movement. All of these hypotheses are at least partially supported by experimental data.

Two VPg precursors from TuMV have been shown to interact with eIF4E and eIF(iso)4E *in planta*; while the VPg-Pro-eIF(iso)4E complex is localized in subnuclear structures, the 6K-VPg-Pro-eIF(iso)4E complex can be detected in cytoplasmic vesicles embedded in the ER (Beauchemin *et al.*, 2007; Leonard *et al.*, 2004). The 6K-VPg-Pro polypeptide has been shown to induce the formation of these cytoplasmic vesicles, which are necessary to assemble the replication complex (Schaad *et al.*, 1997). The 6K-VPg-Pro polypeptide is sufficient on its own to redirect the viral RdRp protein and the host factors PABP, Hsc 70-3, eEF1A, and eIF4E/eIF(iso)4E to these vesicles (Beauchemin and Laliberte, 2007; Beauchemin *et al.*, 2007; Dufresne *et al.*, 2008; Thivierge *et al.*, 2008). These findings suggest that VPg may play an important role in replication and/or translation.

VPg could play a role in the circularization of the viral genome, possibly achieved through an interaction between the VPg-eIF4E/eIF (iso)4E complex and eIF4G/eIF(iso)4G, which itself interacts with PABP, or through the direct interaction between VPg-Pro and PABP (Leonard *et al.*, 2004). Ternary complexes including VPg, eIF4E, and eIF4G (eIF4F) have been identified in several potyvirus infections, including TuMV, LMV, and PVY (Grzela *et al.*, 2006; Michon *et al.*, 2006; Miyoshi *et al.*, 2006). Additionally, *A. thaliana* eIF4G/eIF(iso)4G mutants affect TuMV, LMV, PPV, and ClYVV multiplication (Nicaise *et al.*, 2007) and this correlates with their selective requirement for eIF4E/eIF(iso)4E, indicating that the whole eIF4F/eIF(iso)4F complex may be required for virus multiplication. Surprisingly, mutations in lettuce eIF4E that perturb interactions with eIF4G *in vitro* do not affect the efficiency of LMV infection. Genome circularization therefore may not rely on prior assembly of the eIF4F complex, being mediated by VPg-PABP. Alternatively, mutations might have a limited effect on the *in planta* eIF4E–4G interaction, allowing the eIF4F complex to form.

RNA circularization is important for efficient translation (Kawaguchi and Bailey-Serres, 2002), but it may also be important for replication, as shown for animal picornaviruses, which similarly bear a 5′-linked VPg and 3′ poly(A)-tail (Herold and Andino, 2001). Through this circularization, VPg may be positioned at the 3′-end, favoring its proposed role as a primer for complementary strand synthesis (Puustinen and Makinen,

2004), while the interaction between RdRp and PABP may initiate replication (Dufresne *et al.*, 2008; Wang *et al.*, 2000). The co-localization of RdRp and eIF4E/eIF(iso)4E in the membrane vesicles induced by 6K-VPg-Pro suggests that eIF4E/eIF(iso)4E may also play a role in replication (Thivierge *et al.*, 2008).

In contrast to feline caliciviruses (relatives of potyviruses that also have a 5′-linked VPg and 3′ poly(A) tail) in which the removal of VPg reduces the efficiency of translation (Herbert *et al.*, 1997), it has still not been shown whether VPg is required for potyvirus translation. The calicivirus VPg is thought to act as a cap substitute, promoting translation from VPg-linked viral RNA while inhibiting the translation of capped mRNAs (Chaudhry *et al.*, 2006; Daughenbaugh *et al.*, 2003, 2006; Goodfellow *et al.*, 2005). In contrast, in the absence of TEV (potyvirus) VPg, cap-independent translation has been shown to occur *in vitro* at an internal ribosome entry site (IRES), a process that is eIF4G-dependent but eIF4E-independent (Carrington and Freed, 1990; Gallie, 2001; Niepel and Gallie, 1999). However, more recent results indicate that the VPg-eIF4E/eIF(iso)4E complex may play a direct role in the translation of the potyvirus genome, favoring translation initiation from virus RNA. The TuMV-VPg-eIF4F/eIF(iso)4F complex has more affinity for TEV RNA (containing the TEV-IRES) than eIF4F/eIF(iso)4F alone, suggesting that the complex enhances cap-independent translation by increasing the affinity of eIF4F/eIF(iso)4F for the virus RNA, while inhibiting cap-dependent translation (Khan *et al.*, 2008). The addition of TEV-IRES was shown to enhance the binding affinity between eIF(iso)4F and TuMV-VPg, suggesting that translation initiation involving VPg occurs at the IRES (Khan *et al.*, 2006). In accordance with these data, several VPgs were shown to stimulate translation of uncapped IRES-containing RNAs, while inhibiting cap-dependent translation in wheat germ extract (Grzela *et al.*, 2006; Khan *et al.*, 2008; Miyoshi *et al.*, 2008). These inhibitory effects on *in vitro* translation were shown not to depend on RNase activity (Miyoshi *et al.*, 2008), in contrast to the results obtained in a previous study (Cotton *et al.*, 2006).

The presence of eukaryotic translation initiation and elongation factors together with viral replication proteins suggests that potyvirus RNA translation and replication are coupled (Beauchemin and Laliberte, 2007; Beauchemin *et al.*, 2007; Dufresne *et al.*, 2008; Thivierge *et al.*, 2008). It has been proposed that the VPg-eIF4E/eIF(iso)4E complex could play a role in this process. Translation and replication have been shown to be coupled *in cis* in several viruses, such as BMV (Yi *et al.*, 2007), *Red clover necrotic mosaic virus* (Mizumoto *et al.*, 2006), *Sindbis virus* (Sanz *et al.*, 2007), and *Poliovirus* (Egger and Bienz, 2005; Egger *et al.*, 2000). Since the 6K-VPg-Pro polypeptide is membrane bound, it may act as an anchoring point for the other proteins (Beauchemin and Laliberte, 2007). Therefore, it has been proposed that this polypeptide could have the important role of trapping host replication and translation machineries through its interaction with eIF4E/eIF(iso)4E, and sequestering them in the cytoplasmic vesicles used for viral multiplication. The virus-induced vesicles may provide optimal conditions for RNA and protein synthesis, protecting the products from host nucleases and proteases, as seen for BMV (Schwartz *et al.*, 2002) and *Hepatitis C virus* (Aizaki *et al.*, 2004). TuMV VPg-Pro interacts with eIF(iso)4E in subnuclear structures *in planta* (Beauchemin *et al.*, 2007), indicating that another role of the VPg-eIF4E/eIF(iso)4E complex is the disruption of nuclear functions. This possibly explains the host-gene shutoff observed for potyviruses (Aranda and Maule, 1998).

The inability to detect potyvirus multiplication in most cases of eIF4E/eIF(iso)4E-mediated resistance could be masking a further role of the VPg-eIF4E/eIF(iso)4E complex in virus cell-to-cell movement. The pepper *pvr2*^{1} mediated resistance is thought to block PVY cell-to-cell movement, since virus accumulation occurs in the same way in inoculated protoplasts of susceptible and resistant varieties (Arroyo *et al.*, 1996). Thus, mutations in eIF4E do not necessarily need to prevent virus multiplication in single cells. The role of eIF4E in cell-to-cell movement might be linked to interactions between the cytoskeleton and the translational machinery, resulting from its strong affinity for eIF4G which is known to bind to microtubules (Bokros *et al.*, 1995). Additionally, the viral proteins CI, CP, and VPg have been implicated in the control of cell-to-cell movement (Revers *et al.*, 1999). TuMV VPg-Pro is localized in subnuclear structures *in planta* and can interact with eIF(iso)4E (Beauchemin *et al.*, 2007). This subnuclear localization might play a role in cell-to-cell movement, as described for umbraviruses, where nucleolar localization of the virus transport protein is required for the subsequent formation of viral ribonucleoprotein particles capable of long-distance movement and systemic infection (Kim *et al.*, 2007). Also, the pea *sbm1* resistance gene might affect PSbMV cell-to-cell movement in addition to its effect on multiplication at the single-cell level (Gao *et al.*, 2004). The bombardment of resistant pea plants with plasmids expressing wild-type eIF4E and a GFP-expressing virus was shown to complement virus multiplication in single cells and virus movement from one cell to another, up to three cells beyond the primary target. This observation can be explained by a direct role of eIF4E in cell-to-cell movement, but also by indirect effects. For example, following efficient accumulation of the virus due to the presence of recombinant wild-type eIF4E in resistant cells, the virus would move to the adjacent cells, perhaps dragging eIF4E along with it. Alternatively, the wild type eIF4E might be capable of moving by itself. In this manner, PSbMV multiplication would not only be complemented in the bombarded cells but also in their immediate neighbors. Further experiments would be required to clarify these aspects.

V. RESISTANCE MEDIATED BY EIF4E/EIF(ISO)4E AGAINST NON-POTYVIRUSES

Only two cases of eIF4E-mediated resistance have been described outside the family *Potyviridae*. One involves the cucumovirus CMV, the other the carmovirus MNSV. An EMS-induced mutation in *A. thaliana eIF4E1* (*cum1-1*), resulting in a stop codon at position 99, was shown to reduce CMV multiplication (Yoshii *et al.*, 2004). Whereas CMV RNA and coat protein accumulated to wild-type levels in *cum1-1* protoplasts (Yoshii *et al.*, 1998a), the accumulation of the 3a protein, required for cell-to-cell movement, was strongly reduced (Yoshii *et al.*, 2004). Interestingly, CMV RNA3, which encodes the 3a protein, continued to accumulate as normal. CMV RNAs are capped at their 5′ ends, but have no 3′ poly(A) tail. Comparative studies of CMV RNAs 3 and 4 have highlighted the effect of their 5′- and 3′-UTRs on translation in protoplasts, using the firefly luciferase gene (*luc*) as a reporter. It was shown that their identical 3′-UTRs contained translation enhancing elements, and that a decrease in the translation efficiency was observed in *cum1-1* protoplasts in comparison to wild-type protoplasts for constructs containing the RNA3 5′-UTR as well as the 3′-UTR. The authors concluded that the RNA3 5′-UTR was responsible for the reduced production of the 3a protein observed in *cum1-1* protoplasts, when the 3′-UTR was present *in cis*. (Yoshii *et al.*, 2004).

Whereas the truncation of *A. thaliana* eIF4E affected the translation of CMV RNA3, a single mutation in melon eIF4E is sufficient to affect the translation of the whole MNSV genome. The MNSV genomic and subgenomic RNAs are uncapped and have no poly(A) tail. A His228Leu substitution near the C-*terminus* of melon eIF4E has been shown to confer resistance to MNSV at the single-cell level, affecting all tested MNSV isolates except MNSV-264 (Diaz *et al.*, 2002; Nieto *et al.*, 2006). The virulence determinant of the resistance-breaking strain was localized to the 3′-UTR of the virus genomic RNA (Diaz *et al.*, 2004). Chimeric viruses were constructed containing parts of the virulent (resistance-breaking) strain MNSV-264 and parts of the avirulent strain MNSV-Mα5. These allowed the virulence and avirulence determinants to be mapped to stretches of 49 and 26 nucleotides, respectively.

The translational efficiency of a *luc* reporter gene flanked by 5′- and 3′-UTRs from virulent, avirulent and chimeric MNSV viruses was analyzed *in vitro* using wheat germ extract, and *in vivo* in melon protoplasts. These experiments showed that: (i) the virulence determinant mediated

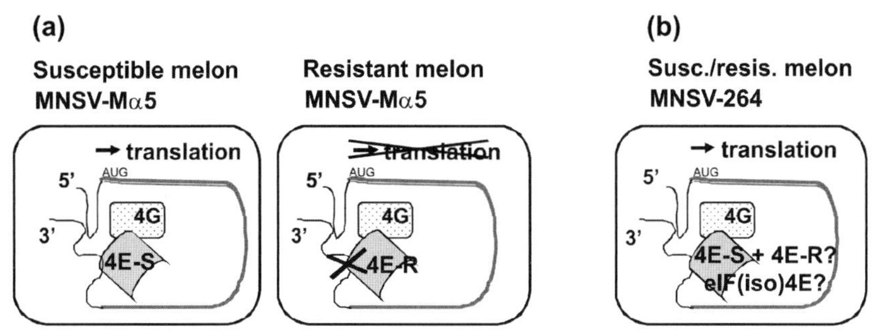

FIGURE 2 Model for the molecular mechanism of the *nsv* resistance. (a) Translation initiation of MNSV-Mα5 in susceptible or resistant melon cells. Efficient interaction between the 3′-CITE and *Cm*-eIF4ES (4E-S) leads to the formation of the translation initiation complex that, through its circularization, ends close to the translation start point allowing efficient translation of the viral RNA. In contrast, inefficient interaction between the 3′-CITE and *Cm*-eIF4ER (4E-R) prevents the formation of the translation initiation complex and translation of the viral RNA. (b) The 3′-CITE of MNSV-264 might interact with both *Cm*-eIF4Es (S and R), or with an eIF4E isoform or might not require eIF4E for its cap-independent translation in susceptible and resistant melon.

efficient cap-independent translation *in vitro* and *in vivo*; (ii) the avirulence determinant was able to promote efficient cap-independent translation *in vitro* but only when eIF4E from susceptible melon (*Cm*-eIF4ES) was added *in trans* and, coherently, only in protoplasts from susceptible melon, not resistant melon; (iii) these activities required the 5′-UTR of MNSV to be present *in cis*. Thus, the virulence and avirulence determinants functioned as cap-independent translation enhancers (3′-CITEs). Since the activity of these 3′-CITEs was host-specific, the *nsv*-mediated resistance must be acting at the level of translation.

Based on these results, a model for the molecular mechanism of *nsv*-mediated resistance to MNSV in melon has been proposed (Fig. 2): circularization of the virus RNA is achieved by base-pairing through a 5′–3′-UTR interaction. In susceptible melon (Fig. 2a), an interaction between the 3′-CITE of MNSV-Mα5 and Cm-eIF4ES allows efficient cap-independent translation initiation, whereas in resistant melon this interaction, and hence translation, is inefficient, preventing viral multiplication in resistant melon cells. On the other hand, as shown in Fig. 2b, the 3′-CITE of MNSV-264 might interact with: (i) both versions of Cm-eIF4E (S and R); (ii) a Cm-eIF4E isoform; or (iii) perhaps another translation initiation factor. The *nsv*-mediated resistance was expected to provide immunity to MNSV (Diaz *et al.*, 2004), but has recently been shown to decline at temperatures below 20 °C (Kido *et al.*, 2008). Thus, the interactions proposed in Fig. 2 may be at least partially temperature-dependent. Alternatively, other processes that occur later in the virus cycle may modulate the effect of cap-independent translation initiation.

VI. RESISTANCE GENES CODING FOR EIF4G OR EIF(ISO)4G

Mutations in other components of the translation initiation complex, such as eIF4G/eIF(iso)4G, have been shown to affect TuMV, LMV, PPV, and ClYVV multiplication in *A. thaliana* (Nicaise *et al.*, 2007) and RYMV multiplication in rice (Albar *et al.*, 2006). In *A. thaliana,* the selective recruitment of these potyviruses for eIF4G/eIF(iso)4G correlates with their selective need for eIF4E/eIF(iso)4E, indicating that the whole eIF4F/eIF(iso)4F complex is required for viral multiplication in single cells. On the other hand, eIF(iso)4G-mediated resistance to RYMV in rice interferes with cell-to-cell movement, whereas virus accumulation in protoplasts from resistant plants is not affected compared to susceptible varieties (Albar *et al.*, 2003). RYMV belongs to the genus *Sobemovirus*, and like potyviruses has a 5′-linked VPg. Common with resistance against potyviruses mediated by eIF4E/eIF(iso)4E, resistance-breaking has been shown to correlate with mutations in the VPg gene (Hébrard *et al.*, 2006). A direct interaction between eIF4G and VPg has been proposed (Hébrard *et al.*, 2008) and the mechanism of this resistance may be similar to the *pvr2*^{1}-mediated resistance against PVY in pepper (Arroyo *et al.*, 1996). As mentioned above, the role of eIF4G (and eIF4E) in cell-to-cell movement may reflect the link between the translational machinery and the cytoskeleton (Bokros *et al.*, 1995). Additionally, interactions have been reported between eIF(iso)4G, VPg-Pro, and 6K-VPg-Pro of TuMV mediated by eIF(iso)4E (Plante *et al.*, 2004), suggesting that eIF(iso)4G could also be localized in subnuclear structures, as found for the TuMV VPg-Pro-eIF(iso)4E complex, playing an indirect role in cell-to-cell movement.

EIF4G-mediated resistance against CMV and TCV (both lack a 5′-linked VPg) has been reported in the *A. thaliana cum2-1* mutant, resulting from a single amino acid substitution (Yoshii *et al.*, 1998b, 2004). The effect of the *cum2-1* mutation on CMV multiplication is similar to that of *cum1-1* (see previous section): CMV RNA and coat protein accumulated to wild-type levels in protoplasts, while the accumulation of the 3a protein, required for cell-to-cell movement, was strongly reduced (Yoshii *et al.*, 2004). As for the *cum1-1* mutant discussed above, the 5′-UTR of RNA3 was found to be responsible for the decrease in 3a protein production in *cum2-1* protoplasts, explaining the effect on cell-to-cell movement and the delay in CMV multiplication. In contrast, TCV genomic RNA accumulation was reduced in *cum2-1* protoplasts compared to wild-type. TCV, like MNSV, belongs to the genus *Carmovirus*, and the genomic RNA has no 5′-cap and no 3′ poly(A)-tail. Comparative studies using the *luc* reporter gene to determine how the TCV 5′- and 3′-UTRs affect translation in protoplasts, showed that the 3′-UTR contained a 3′-CITE and that the 5′- and 3′-UTRs synergistically enhanced translation. Translation efficiency controlled by both UTRs was found to be reduced in *cum2-1* protoplasts compared to wild-type, thus pinpointing the resistance mechanism. The fact that TCV translation in *cum2-1* protoplasts was inhibited to a lesser degree than MNSV translation in resistant melon protoplasts, correlates with the reduced TCV multiplication compared to the lack of MNSV accumulation observed in the corresponding resistant protoplasts.

VII. CONCLUSIONS AND PROSPECTS

Over the last few years, a small but significant proportion of the natural recessive resistance genes listed by Kang and colleagues (Kang *et al.*, 2005b) have been cloned and sequenced. Strikingly, all of them have been shown to encode translation initiation factors of the 4E and 4G families. Does this mean that all recessive resistance genes encode these proteins? This is probably not the case. First, a large number of natural resistance genes have yet to be characterized, so much remains to be learned about the nature of recessive resistance. Second, screens of mutagenized host populations have found susceptibility factors other than eIF4E or eIF4G, hinting that different mechanisms exist. Finally, we are aware of at least two cases in which *eIF4E* and/or *eIF4G* were unsuccessfully used as candidate genes—the cucumber *zym* and melon *cmv1* genes, indicating these genes encode distinct proteins (Essafi *et al.*, 2009; Meyer *et al.*, 2008). Although genes belonging to the 4E and 4G families are not the only possible recessive resistance genes, they do show up very frequently. We can identify two main reasons for this. First, there are several isoforms of these translation initiation factors in all plant species and there seems to be some functional redundancy among them which provides scope for diverse interactions; and second, interactions between these factors and virus proteins can be quite specific, providing niches for individual viruses. These two conditions may have helped to isolate eIF4E and eIF4G from other susceptibility factors and provided the flexibility required for the resistance/resistance-breaking evolutionary arms race.

Given the cellular role of eIF4E and eIF4G in translation initiation, a straightforward explanation of their role in the virus replication cycle would be their participation in the translation initiation of viral RNAs. This has been confirmed in the case of eIF4E/eIF4G-mediated resistances against viruses lacking VPg, such are MNSV (melon *nsv* resistance) (Truniger *et al.*, 2008) and CMV and TCV (*A. thaliana cum-1* and *cum-2*) (Yoshii *et al.*, 2004). However, the mechanism(s) underlying eIF4E-mediated resistance to potyviruses are far from clear. In the potyvirus cases studied thus far, a physical interaction between VPg and eIF4E is necessary for virus multiplication but the precise role of this interaction has yet to be determined. In order to understand this, more information needs to be collected on the early steps of the potyvirus infection cycle. For example, we are not aware of any report looking at whether or not the translatable form of the potyvirus RNA is linked to VPg *in vivo*. The dual subcellular localization of the VPg-eIF4E complex is particularly intriguing—eIF4E is recruited to cytoplasmic vesicles by VPg where viral translation and replication occur, but what is the purpose of a VPg–eIF4E interaction in the nucleus? In this regard, a series of observations throughout the potyvirus cellular multiplication cycle could provide insight into the phenomenon and characterize how interactions change as the virus progresses through its infection cycle.

As discussed above, the cloning, sequencing, and characterization of recessive resistance genes is helping to provide fundamental information on how viruses interact with their hosts to achieve multiplication, and how hosts are reacting to prevent it. The increasing availability of genomic information for crop species and new, powerful molecular biology tools (e.g., technologies for high-throughput sequencing, expression profiling, and proteomics) allows the application of novel techniques to this problem, such as high-throughput interaction screens to characterize plant–virus interactomes, and screens of mutagenized host populations to rapidly identify novel mutants. Regarding the characterization of plant–virus interactomes, it is worth remembering that the early identification of an interaction between potyvirus VPg and eIF(iso)4E (Wittmann *et al.*, 1997) pointed rapidly toward eIF4E/iso4E as a candidate host susceptibility factor. Regarding screens of mutagenized hosts, the cloning and engineering of viral genomes allows the creation of labeled virus strains that facilitate screening and phenotypic analysis. For example, early work by Whitham and colleagues (1999) led to the development of a selectable TEV strain which expressed a herbicide resistance gene to facilitate the efficient identification of either gain-of-susceptibility or loss-of-susceptibility mutants in a large collection of mutagenized *A. thaliana* plants.

From a strategic point of view, research in this area should provide new sources of resistance against plant pathogenic viruses. The translation of current knowledge from model to crop species may contribute to this objective. Genetic screens for alleles defective in conferring susceptibility require the accumulation of genomic data for important crop species, and the generation of platforms for the efficient genetic analysis of large collections of mutagenized lines, example by TILLING (Targeted Induced Local Lesions IN Genomes (Oleykowski *et al.*, 1998; Slade *et al.*, 2005) or ecotypes, for example by EcoTILLING (Nieto *et al.*, 2007). Whether or not such new sources of resistance will be durable can only be confirmed by testing in the field, although the study of the evolution of viruses, and of the processes involved in the emergence and reemergence of viral diseases, have the potential to identify useful resistance durability predictors. At the very least, recessive resistance genes and viruses have already provided fundamental information on host–virus coevolution (Cavatorta *et al.*, 2008; Charron *et al.*, 2008; Moury *et al.*, 2004; Sacristan and Garcia-Arenal, 2008).

REFERENCES

Abdul-Razzak, A., Guiraud, T., Peypelut, M., Walter, J., Houvenaghel, M. C., Candresse, T., Le Gall, O., and German-Retana, S. (2009). Involvement of the cylindrical inclusion (CI) protein in the overcoming of an eIF4E-mediated resistance against Lettuce mosaic potyvirus. *Mol. Plant Pathol.* **10**:109–113.

Aizaki, H., Lee, K.-J., Sung, V. M. H., Ishiko, H., and Lai, M. M. C. (2004). Characterization of the hepatitis C virus RNA replication complex associated with lipid rafts. *Virology* **324**:450–461.

Albar, L., Ndjiondjop, M. N., Esshak, Z., Berger, A., Pinel, A., Jones, M., Fargette, D., and Ghesquière, A. (2003). Fine genetic mapping of a gene required for Rice yellow mottle virus cell-to-cell movement. *Theor. Appl. Genet.* **107**:371–378.

Albar, L., Bangratz-Reyser, M., Hebrard, E., Ndjiondjop, M. N., Jones, M., and Ghesquiere, A. (2006). Mutations in the eIF(iso)4G translation initiation factor confer high resistance of rice to Rice yellow mottle virus. *Plant J.* **47**:417–426.

Aranda, M., and Maule, A. (1998). Virus-Induced Host Gene Shutoff in Animals and Plants. *Virology* **243**:261–267.

Arroyo, R., Soto, M., Martínez-Zapater, J., and Ponz, F. (1996). Impaired cell-to-cell movement of Potato Virus Y in pepper plants carrying the y^a ($pr2^1$) resistance gene. *Mol. Plant-Microbe Interact.* **9**:314–318.

Asano, M., Satoh, R., Mochizuki, A., Tsuda, S., Yamanaka, T., Nishiguchi, M., Hirai, K., Meshi, T., Naito, S., and Ishikawa, M. (2005). Tobamovirus-resistant tobacco generated by RNA interference directed against host genes. *FEBS Lett.* **579**:4479–4484.

Beauchemin, C., Boutet, N., and Laliberte, J. F. (2007). Visualization of the interaction between the precursors of VPg, the viral protein linked to the genome of turnip mosaic virus, and the translation eukaryotic initiation factor iso4E *in planta*. *J. Virol.* **81**:775–782.

Beauchemin, C., and Laliberte, J.-F. (2007). The poly(A) binding protein is internalized in virus-induced vesicles or redistributed to the nucleolus during turnip mosaic virus infection. *J. Virol.* **81:**10905–10913.

Boiteux, L. S., Cupertino, F. P., Silva, C., Dusi, A. N., MonteNeshich, D. C., vanderVlugt, R. A. A., and Fonseca, M. E. N. (1996). Resistance to potato virus Y (pathotype 1-2) in *Capsicum annuum* and *Capsicum chinense* is controlled by two independent major genes. *Euphytica* **87:**53–58.

Bokros, C. L., Hugdahl, J. D., Kim, H., Hanesworth, V. R., Heerden, A. V., Browning, K. S., and Morejohn, L. C. (1995). Function of the p86 subunit of eukaryotic initiation factor (iso)4F as a microtubule-associated protein in plant cells. *Proc. Natl. Acad. Sci. U. S. A.* **92:**7120–7124.

Borgstrom, B., and Johansen, I. E. (2001). Mutations in *Pea seedborne mosaic virus* genome-linked protein VPg alter pathotype-specific virulence in *Pisum sativum*. *Mol. Plant-Microbe Interact.* **14:**707–714.

Browning, K. S. (1996). The plant translational apparatus. *Plant Mol. Biol.* **32:**107–144.

Browning, K. S. (2004). Plant translation initiation factors: it is not easy to be green. *Biochem. Soc. Trans.* **32:**589–591.

Bruun-Rasmussen, M., Moller, I. S., Tulinius, G., Hansen, J. K. R., Lund, O. S., and Johansen, I. E. (2007). The same allele of translation initiation factor 4E mediates resistance against two *Potyvirus* spp. in *Pisum sativum*. *Mol. Plant-Microbe Interact.* **20:**1075–1082.

Buschges, R., Hollricher, K., Panstruga, R., and Schulze-Lefert, P. (1997). The barley *Mlo* gene: a novel control element of plant pathogen resistance. *Cell* **88:**695–705.

Candresse, T., Le Gall, O., Maisonneuve, B., German-Retana, S., and Redondo, E. (2002). The use of green fluorescent protein-tagged recombinant viruses to test lettuce mosaic virus resistance in lettuce. *Phytopathology* **92:**169–176.

Caranta, C., Palloix, A., GebreSelassie, K., Lefebvre, V., Moury, B., and Daubeze, A. M. (1996). A complementation of two genes originating from susceptible *Capsicum annuum* lines confers a new and complete resistance to pepper veinal mottle virus. *Phytopathology* **86:**739–743.

Carrington, J. C., and Freed, D. D. (1990). Cap-independent enhancement of translation by a plant potyvirus 5′ nontranslated region. *J. Virol.* **64:**1590–1597.

Carrington, J. C., Freed, D. D., and Leinicke, A. J. (1991). Bipartite signal sequence mediates nuclear translocation of the plant potyviral Nla protein. *Plant Cell* **3:**953–962.

Carrington, J., Jensen, P., and Schaad, M. C. (1998). Genetic evidence for an essential role for potyvirus CI protein in cell to cell movement. *Plant J.* **14:**393–400.

Cavatorta, J., Savage, A., Yeam, I., Gray, S., and Jahn, M. (2008). Positive Darwinian selection at single amino acid sites conferring plant virus resistance. *J. Mol. Evol.* **67:**551–559.

Charron, C., Nicolaï, M., Gallois, J.-L., Robaglia, C., Moury, B., Palloix, A., and Caranta, C. (2008). Natural variation and functional analyses provide evidence for co-evolution between plant eIF4E and potyviral VPg. *Plant J.* **54:**56–68.

Chaudhry, Y., Nayak, A., Bordeleau, M.-E., Tanaka, J., Pelletier, J., Belsham, G. J., Roberts, L. O., and Goodfellow, I. G. (2006). Caliciviruses differ in their functional requirements for eIF4F components. *J. Biol. Chem.* **281:**25315–25325.

Cotton, S., Dufresne, P. J., Thivierge, K., Ide, C., and Fortin, M. G. (2006). The VPgPro protein of *Turnip mosaic virus*: *In vitro* inhibition of translation from a ribonuclease activity. *Virology* **351:**92–100.

Daughenbaugh, K., Fraser, C., Hershley, J., and Hardy, M. (2003). The genome-linked protein Vpg of the Norwalk virus binds eIF3, suggesting its role in translation initiation complex recruitment. *EMBO J.* **22:**2852–2859.

Daughenbaugh, K., Wobus, C., and Hardy, M. (2006). VPg of murine norovirus binds translation initiation factors in infected cells. *Virology J.* **3**

Decroocq, V., Sicard, O., Alamillo, J. M., Lansac, M., Eyquard, J. P., García, J., Candresse, T., Le Gall, O., and Revers, F. (2006). Multiple resistance traits control Plum pox virus infection in *Arabidopsis thaliana*. *Mol. Plant-Microbe Interact.* **19:**541–549.

Deom, C. M., Murphy, J. F., and Paguio, O. R. (1997). Resistance to tobacco etch virus in *Capsicum annuum*: Inhibition of virus RNA accumulation. *Mol. Plant-Microbe Interact.* **10:**917–921.

Diaz, J. A., Nieto, C., Moriones, E., and Aranda, M. A. (2002). Spanish *Melon necrotic spot virus* isolate overcomes the resistance conferred by the recessive nsv gene of melon. *Plant Dis.* **86:**694.

Diaz, J. A., Nieto, C., Moriones, E., Truniger, V., and Aranda, M. A. (2004). Molecular Characterization of a *Melon necrotic spot virus* strain that overcomes the resistance in melon and nonhost plants. *Mol. Plant-Microbe Interact.* **17:**668–675.

Díaz-Pendón, J. A., Truniger, V., Nieto, C., Garcia-Mas, J., Bendahmane, A., and Aranda, M. A. (2004). Advances in understanding recessive resistance to plant viruses. *Mol. Plant Pathol.* **5:**223–233.

Dogimont, C., Palloix, A., Daubze, A. M., Marchoux, G., Selassie, K. G., and Pochard, E. (1996). Genetic analysis of broad spectrum resistance to potyviruses using doubled haploid lines of pepper (*Capsicum annuum L*). *Euphytica* **88:**221–239.

Dufresne, P. J., Thivierge, K., Cotton, S., Beauchemin, C., Ide, C., Ubalijoro, E., Laliberté, J.-F., and Fortin, M. G. (2008). Heat shock 70 protein interaction with *Turnip mosaic virus* RNA-dependent RNA polymerase within virus-induced membrane vesicles. *Virology* **374:**217–227.

Duprat, A., Caranta, C., Revers, F., Menand, B., Browning, K. S., and Robaglia, C. (2002). The Arabidopsis eukaryotic initiation factor (iso)4E is dispensable for plant growth but required for susceptibility to potyviruses. *Plant J.* **32:**927–934.

Egger, D., and Bienz, K. (2005). Intracellular location and translocation of silent and active poliovirus replication complexes. *J. Gen. Virol.* **86:**707–718.

Egger, D., Teterina, N., Ehrenfeld, E., and Bienz, K. (2000). Formation of the poliovirus replication complex requires coupled viral translation, vesicle production, and viral RNA synthesis. *J. Virol.* **74:**6570–6580.

Essafi, A., Díaz-Pendón, J., Moriones, E., Monforte, A., Garcia-Mas, J., and Martín-Hernández, A. (2009). Dissection of the oligogenic resistance to *Cucumber mosaic virus* in the melon accession PI 161375. *Theor. Appl. Genet.* **118:**275–284.

Fellers, J., Wan, J., Hong, Y., Collins, G. B., and Hunt, A. G. (1998). In vitro interactions between a potyvirus-encoded, genome-linked protein and RNA-dependent RNA polymerase. *J. Gen. Virol.* **79:**2043–2049.

Fernandez, A., Guo, H. S., Saenz, P., Simon-Buela, L., Gomez de Cedron, M., and Garcia, J. A. (1997). The motif V of plum pox potyvirus CI RNA helicase is involved in NTP hydrolysis and is essential for virus RNA replication. *Nucl. Acids Res.* **25:**4474–4480.

Fraser, R. S. S. (1990). The genetics of resistance to plant viruses. *Annu. Rev. Phytopathol.* **28:**179–200.

Gallie, D. R. (2001). Cap-independent translation conferred by the 5′ leader of tobacco etch virus is eukaryotic initiation factor 4G dependent. *J. Virol.* **75:**12141–12152.

Gallie, D. R., and Browning, K. S. (2001). eIF4G functionally differs from eIFiso4G in promoting internal initiation, cap-independent translation, and translation of structured mRNAs. *J. Biol. Chem.* **276:**36951–36960.

Gao, Z., Johansen, E., Eyers, S., Thomas, C. L., Noel Ellis, T. H., and Maule, A. J. (2004). The potyvirus recessive resistance gene, *sbm1*, identifies a novel role for translation initiation factor eIF4E in cell-to-cell trafficking. *Plant J.* **40:**376–385.

German-Retana, S., Candresse, T., Alias, E., Delbos, R.-P., and Le Gall, O. (2000). Effects of green fluorescent protein or ß-glucuronidase tagging on the accumulation and pathogenicity of a resistance-breaking Lettuce mosaic virus Isolate in susceptible and resistant lettuce cultivars. *Mol. Plant-Microbe Interact.* **13:**316–324.

German-Retana, S., Walter, J., Doublet, B., Roudet-Tavert, G., Nicaise, V., Lecampion, C., Houvenaghel, M.-C., Robaglia, C., Michon, T., and Le Gall, O. (2008). Mutational analysis of plant cap-binding protein eIF4E reveals key amino acids involved in biochemical functions and potyvirus infection. *J. Virol.* **82:**7601–7612.

Gómez de Cedrón, M., Osaba, L., López, L., and García, J. A. (2006). Genetic analysis of the function of the plum pox virus CI RNA helicase in virus movement. *Virus Res.* **116:**136–145.

Goodfellow, I. G., and Roberts, L. O. (2008). Eukaryotic initiation factor 4E. *Int. J. Biochem. & Cell Biol.* **40:**2675–2680.

Goodfellow, I. G., Chaudhry, Y., Gioldasi, I., Gerondopoulus, A., Natoni, A., lLabrie, L., Laliberté, J., and Roberts, L. O. (2005). Calicivirus translation initiation requires an interaction between Vpg and eIF4E. *EMBO Rep.* **6:** 968–972.

Grzela, R., Strokovska, L., Andrieu, J. P., Dublet, B., Zagorski, W., and Chroboczek, J. (2006). Potyvirus terminal protein VPg, effector of host eukaryotic initiation factor eIF4E. *Biochimie* **88:**887–896.

Grzela, R., Szolajska, E., Ebel, C., Madern, D., Favier, A., Wojtal, I., Zagorski, W., and Chroboczek, J. (2008). Virulence factor of potato virus Y, genome-attached terminal protein VPg, is a highly disordered protein. *J. Biol. Chem.* **283:** 213–221.

Hagiwara, Y., Komoda, K., Yamanaka, T., Tamai, A., Meshi, T., Funada, R., Tsuchiya, T., Naito, S., and Ishikawa, M. (2003). Subcellular localization of host and viral proteins associated with tobamovirus RNA replication. *EMBO J.* **22:**344–353.

Hébrard, E., Pinel-Galzi, A., Bersoult, A., Sire, C., and Fargette, D. (2006). Emergence of a resistance-breaking isolate of *Rice yellow mottle virus* during serial inoculations is due to a single substitution in the genome-linked viral protein VPg. *J. Gen. Virol.* **87:**1369–1373.

Hébrard, E., Pinel-Galzi, A., and Fargette, D. (2008). Virulence domain of the RYMV genome-linked viral protein VPg towards rice *rymv 1–2*-mediated resistance. *Arch. Virol.* **153:**1161–1164.

Herbert, T. P., Brierley, I., and Brown, T. D. (1997). Identification of a protein linked to the genomic and subgenomic mRNAs of feline calicivirus and its role in translation. *J. Gen. Virol.* **78:**1033–1040.

Herold, J., and Andino, R. (2001). Poliovirus RNA replication requires genome circularization through a protein-protein bridge. *Mol. Cell* **7:**581–591.

Hong, Y., Levay, K., Murphy, J. F., Klein, P. G., Shaw, J. G., and Hunt, A. G. (1995). A potyvirus polymerase interacts with the viral coat protein and VPg in yeast cells. *Virology* **214:**159–166.

Ishikawa, M., Obata, F., Kumagai, T., and Ohno, T. (1991). Isolation of mutants of Arabidopsis thaliana in which accumulation of tobacco mosaic virus coat protein is reduced to low levels. *Mol. Gen. Genet.* **230:**33–38.

Ishikawa, M., Naito, S., and Ohno, T. (1993). Effects of the *tom1* mutation of *Arabidopsis thaliana* on the multiplication of tobacco mosaic virus RNA in protoplasts. *J. Virol.* **67:**5328–5338.

Janda, M., and Ahlquist, P. (1993). RNA-dependent replication, transcription, and persistence of brome mosaic virus RNA replicons in *S. cerevisiae*. *Cell* **72:**961–970.

Johansen, I. E., Lund, O. S., Hjulsager, C. K., and Laursen, J. (2001). Recessive resistance in *Pisum sativum* and *Potyvirus* pathotype resolved in a gene-for-cistron correspondence between host and virus. *J. Virol.* **75:**6609–6614.

Kang, B.-C., Yeam, I., Frantz, J. D., Murphy, J. F., and Jahn, M. M. (2005). The *pvr1* locus in *Capsicum* encodes a translation initiation factor eIF4E that interacts with *Tobacco etch virus* VPg. *Plant J.* **42:**392–405.

Kang, B.-C., Yeam, I., and Jahn, M. M. (2005). Genetics of Plant Virus Resistance. *Annu. Rev. Phytopathol.* **43:**581–621.

Kang, B.-C., Yeam, I., Li, H. X., Perez, K. W., and Jahn, M. M. (2007). Ectopic expression of a recessive resistance gene generates dominant potyvirus resistance in plants. *Plant Biotech. J.* **5:**526–536.

Kanyuka, K., McGrann, G., Alhudaib, K., Hariri, D., and Adams, M. J. (2004). Biological and sequence analysis of a novel European isolate of *Barley mild mosaic virus* that overcomes the barley *rym5* resistance gene. *Arch. Virol.* **149:**1469–1480.

Kanyuka, K., Druka, A., Caldwell, D., Tymon, A., McCallum, N., Waugh, R., and Adams, M. J. (2005). Evidence that the recessive bymovirus resistance locus *rym4* in barley corresponds to the eukaryotic translation initiation factor 4E gene. *Mol. Plant Pathol.* **6**:449–458.

Kawaguchi, R., and Bailey-Serres, J. (2002). Regulation of translational initiation in plants. *Curr. Op. Plant Biol.* **5**:460–465.

Kekarainen, T., Savilahti, H., and Valkonen, J. P. T. (2002). Functional genomics on Potato Virus A: virus genome-wide map of sites essential for virus propagation. *Genome Res.* **12**:584–594.

Keller, K. E., Johansen, E., Martin, R. R., and Hampton, R. O. (1998). Potyvirus genome-linked protein (VPg) determines pea seed-borne mosaic virus pathotype-specific virulence in Pisum sativum. *Mol. Plant-Microbe Interact.* **11**:124–130.

Khan, M. A., Miyoshi, H., Ray, S., Natsuaki, T., Suehiro, N., and Goss, D. J. (2006). Interaction of genome-linked Protein (VPg) of turnip mosaic virus with wheat germ translation initiation factors eIFiso4E and eIFiso4F. *J. Biol. Chem.* **281**:28002–28010.

Khan, M. A., Miyoshi, H., Gallie, D. R., and Goss, D. J. (2008). Potyvirus genome-linked protein, VPg, directly affects wheat germ *in vitro* translation: interactions with translation initiation factors eIF4F and eIFiso4F. *J. Biol. Chem.* **283**:1340–1349.

Khraiwesh, B., Ossowski, S., Weigel, D., Reski, R., and Frank, W. (2008). Specific gene silencing by artificial microRNAs in *Physcomitrella patens*: An alternative to targeted gene knockouts. *Plant Physiol.* **148**:684–693.

Kido, K., Mochizuki, T., Matsuo, K., Tanaka, C., Kubota, K., Ohki, T., and Tsuda, S. (2008). Functional degeneration of the resistance gene *nsv* against *Melon necrotic spot virus* at low temperature. *Europ. J. Plant Pathol.* **121**:189–194.

Kim, S. H., Ryabov, E. V., Kalinina, N. O., Rakitina, D., Gillespie, T., Macfarlane, S., Haupt, S., Brown, J., and Taliansky, M. E. (2007). Cajal bodies and the nucleolus are required for a plant virus systemic infection. *EMBO J.* **26**:2169–2179.

Kuhne, T., Shi, N., Proeseler, G., Adams, M. J., and Kanyuka, K. (2003). The ability of a bymovirus to overcome the *rym4*-mediated resistance in barley correlates with a codon change in the VPg coding region on RNA1. *J. Gen. Virol.* **84**:2853–2859.

Kushner, D. B., Lindenbach, B. D., Grdzelishvili, V. Z., Noueiry, A. O., Paul, S. M., and Ahlquist, P. (2003). Systematic, genome-wide identification of host genes affecting replication of a positive-strand RNA virus. *Proc. Natl. Acad. Sci. U. S. A.* **100**:15764–15769.

Kyle, M. M., and Palloix, A. (1997). Proposed revision of nomenclature for potyvirus resistance genes in Capsicum. *Euphytica* **97**:183–188.

Lellis, A. D., Kasschau, K. D., Whitham, S. A., and Carrington, J. C. (2002). Loss-of-susceptibility mutants of *Arabidopsis thaliana* reveal an essential role for eIF (iso)4E during potyvirus infection. *Curr. Biol.* **12**:1046–1051.

Leonard, S., Plante, D., Wittmann, S., Daigneault, N., Fortin, M. G., and Laliberte, J.-F. (2000). Complex formation between potyvirus VPg and translation eukaryotic initiation factor 4E correlates with virus infectivity. *J. Virol.* **74:** 7730–7737.

Leonard, S., Viel, C., Beauchemin, C., Daigneault, N., Fortin, M. G., and Laliberte, J. F. (2004). Interaction of VPg-Pro of *Turnip mosaic virus* with the translation

initiation factor 4E and the poly(A)-binding protein *in planta*. *J. Gen. Virol.* **85:**1055–1063.
Meyer, J., Deleu, W., Garcia-Mas, J., and Havey, M. (2008). Construction of a fosmid library of cucumber (*Cucumis sativus*) and comparative analyses of the eIF4E and eIF(iso)4E regions from cucumber and melon (*Cucumis melo*). *Mol. Genet. Genomics* **279:**473–480.
Michon, T., Estevez, Y., Walter, J., German-Retana, S., and Le Gall, O. (2006). The potyviral virus genome-linked protein VPg forms a ternary complex with the eukaryotic initiation factors eIF4E and eIF4G and reduces eIF4E affinity for a mRNA cap analogue. *FEBS J.* **273:**1312–1322.
Miyoshi, H., Suehiro, N., Tomoo, K., Muto, S., Takahashi, T., Tsukamoto, T., Ohmori, T., and Natsuaki, T. (2006). Binding analyses for the interaction between plant virus genome-linked protein (VPg) and plant translational initiation factors. *Biochimie* **88:**329–340.
Miyoshi, H., Okade, H., Muto, S., Suehiro, N., Nakashima, H., Tomoo, K., and Natsuaki, T. (2008). Turnip mosaic virus VPg interacts with Arabidopsis thaliana eIF(iso)4E and inhibits in vitro translation. *Biochimie* **90:**1427–1434.
Mizumoto, H., Iwakawa, H.-O., Kaido, M., Mise, K., and Okuno, T. (2006). Cap-Independent translation mechanism of *Red clover necrotic mosaic virus* RNA2 differs from that of RNA1 and is linked to RNA replication. *J. Virol.* **80:**3781–3791.
Monzingo, A. F., Dhaliwal, S., Dutt-Chaudhuri, A., Lyon, A., Sadow, J. H., Hoffman, D. W., Robertus, J. D., and Browning, K. S. (2007). The structure of eukaryotic translation initiation factor-4E from wheat reveals a novel disulfide bond. *Plant Physiol.* **143:**1504–1518.
Moury, B., Morel, C., Johansen, E., Guilbaud, L., Souche, S., Ayme, V., Caranta, C., Palloix, A., and Jacquemond, M. (2004). Mutations in *Potato virus Y* genome-linked protein determine virulence toward recessive resistances in *Capsicum annuum* and *Lycopersicon hirsutum*. *Mol. Plant-Microbe Interact.* **17:**322–329.
Murphy, J. F., and Kyle, M. M. (1995). Alleviation of restricted systemic spread of pepper mottle potyvirus in *Capsicum annuum* cv. Avelar by coinfection with a cucumovirus. *Phytopathology* **85:**561–566.
Murphy, J. F., Klein, P. G., Hunt, A. G., and Shaw, J. G. (1996). Replacement of the tyrosine residue that links a potyviral VPg to the viral RNA is lethal. *Virology* **220:**535–538.
Murphy, J. F., Blauth, J. R., Livingstone, K. D., Lackney, V. K., and Jahn, M. K. (1998). Genetic mapping of the *pvr1* locus in *Capsicum* spp, and evidence that distinct potyvirus resistance loci control responses that differ at the whole plant and cellular levels. *Mol. Plant-Microbe Interact.* **11:**943–951.
Nagy, P. D. (2008). Yeast as a model host to explore plant virus-host interactions. *Annu. Rev. Phytopathol.* **46:**217–242.
Nagy, P. D., and Pogany, J. (2006). Yeast as a model host to dissect functions of viral and host factors in tombusvirus replication. *Virology* **344:**211–220.
Nicaise, V., German-Retana, S., Sanjuan, R., Dubrana, M.-P., Mazier, M., Maisonneuve, B., Candresse, T., Caranta, C., and Le Gall, O. (2003). The eukaryotic

translation initiation factor 4E controls lettuce susceptibility to the *Potyvirus Lettuce mosaic virus*. *Plant Physiol.* **132:**1272–1282.

Nicaise, V., Gallois, J. L., Chafiai, F., Allen, L. M., Schurdi-Levraud, V., Browning, K. S., Candresse, T., Caranta, C., Le Gall, O., and German-Retana, S. (2007). Coordinated and selective recruitment of eIF4E and eIF4G factors for potyvirus infection in Arabidopsis thaliana. *FEBS Lett.* **581:**1041–1046.

Niepel, M., and Gallie, D. R. (1999). Identification and characterization of the functional elements within the tobacco etch virus 5′ leader required for cap-independent translation. *J. Virol.* **73:**9080–9088.

Nieto, C., Morales, M., Orjeda, G., Clepet, C., Monfort, A., Sturbois, B., Puigdomenech, P., Pitrat, M., Caboche, M., Dogimont, C., Garcia-Mas, J., Aranda, M. A., and Bendahmane, A. (2006). An eIF4E allele confers resistance to an uncapped and non-polyadenylated RNA virus in melon. *Plant J.* **48:**452–462.

Nieto, C., Piron, F., Dalmais, M., Marco, C., Moriones, E., Gomez-Guillamon, M. L., Truniger, V., Gomez, P., Garcia-Mas, J., Aranda, M., and Bendahmane, A. (2007). EcoTILLING for the identification of allelic variants of melon eIF4E, a factor that controls virus susceptibility. *BMC Plant Biol.* **7:**34.

Ohshima, K., Taniyama, T., Yamanaka, T., Ishikawa, M., and Naito, S. (1998). Isolation of a mutant of Arabidopsis thaliana carrying two simultaneous mutations affecting tobacco mosaic virus multiplication within a single cell. *Virology* **243:**472–481.

Oleykowski, C. A., Bronson Mullins, C. R., Godwin, A. K., and Yeung, A. T. (1998). Mutation detection using a novel plant endonuclease. *Nucl. Acids Res.* **26:**4597–4602.

Panavas, T., and Nagy, P. D. (2005). Mechanism of stimulation of plus-strand synthesis by an RNA replication enhancer in a Tombusvirus. *J. Virol.* **79:**9777–9785.

Parrella, G., Ruffel, S., Moretti, A., Morel, C., Palloix, A., and Caranta, C. (2002). Recessive resistance genes against potyviruses are localized in colinear genomic regions of the tomato (*Lycopersicon* spp.) and pepper (*Capsicum* spp.) genomes. *Theor. Appl. Genet.* **105:**855–861.

Plante, D., Viel, C., Leonard, S., Tampo, H., Laliberte, J. F., and Fortin, M. G. (2004). Turnip mosaic virus Vpg does not disrupt the translation initiation complex but interferes with cap binding. *Physiol. Mol. Plant Pathol.* **64:**219–226.

Provvidenti, R., and Hampton, R. O. (1993). Inheritance of resistance to white lupin mosaic virus in common pea. *HortScience* **28:**836–837.

Puustinen, P., and Makinen, K. (2004). Uridylylation of the potyvirus VPg by viral replicase NIb correlates with the nucleotide binding capacity of VPg. *J. Biol. Chem.* **279:**38103–38110.

Rajamäki, M.-L., and Valkonen, J. P. T. (2002). Viral genome-linked protein (VPg) controls accumulation and phloem-loading of a potyvirus in inoculated potato leaves. *Mol. Plant-Microbe Interact.* **15:**138–149.

Rantalainen, K. I., Uversky, V. N., Permi, P., Kalkkinen, N., Dunker, A. K., and Mäkinen, K. (2008). Potato virus A genome-linked protein VPg is an intrinsically disordered molten globule-like protein with a hydrophobic core. *Virology* **377:**280–288.

Revers, F., Lot, H., Souche, S., Le Gall, O., Candresse, T., and Dunez, J. (1997). Biological and molecular variability of lettuce mosaic virus isolates. *Phytopathology* **87**:397–403.

Revers, F. D. R., Le Gall, O., Candresse, T., and Maule, A. J. (1999). New advances in understanding the molecular biology of plant/potyvirus interactions. *Mol. Plant-Microbe Interact.* **12**:367–376.

Robaglia, C., and Caranta, C. (2006). Translation initiation factors: A weak link in plant RNA virus infection. *Trends Plant Sci.* **11**:40–45.

Roudet-Tavert, G., Michon, T., Walter, J., Delaunay, T., Redondo, E., and Le Gall, O. (2007). Central domain of a potyvirus VPg is involved in the interaction with the host translation initiation factor eIF4E and the viral protein HcPro. *J. Gen. Virol.* **88**:1029–1033.

Ruffel, S., Dussault, M. H., Palloix, A., Moury, B., Bendahmane, A., Robaglia, C., and Caranta, C. (2002). A natural recessive resistance gene against potato virus Y in pepper corresponds to the eukaryotic initiation factor 4E (eIF4E). *Plant J.* **32**:1067–1075.

Ruffel, S., Gallois, J., Lesage, M., and Caranta, C. (2005). The recessive potyvirus resistance gene pot-1 is the tomato orthologue of the pepper pvr2-eIF4E gene. *Mol. Genet. Genomics* **274**:346–353.

Ruffel, S., Gallois, J. L., Moury, B., Robaglia, C., Palloix, A., and Caranta, C. (2006). Simultaneous mutations in translation initiation factors elF4E and elF(iso)4E are required to prevent pepper veinal mottle virus infection of pepper. *J. Gen. Virol.* **87**:2089–2098.

Sacristan, S., and Garcia-Arenal, F. (2008). The evolution of virulence and pathogenicity in plant pathogen populations. *Mol. Plant Pathol.* **9**:369–384.

Sanz, M. A., Castello, A., and Carrasco, L. (2007). Viral translation is coupled to transcription in sindbis virus-infected cells. *J. Virol.* **81**:7061–7068.

Sato, M., Masuta, C., and Uyeda, I. (2003). Natural resistance to *Clover yellow vein virus* in beans controlled by a single recessive locus. *Mol. Plant-Microbe Interact.* **16**:994–1002.

Sato, M., Nakahara, K., Yoshii, M., Ishikawa, M., and Uyeda, I. (2005). Selective involvement of members of the eukaryotic initiation factor 4E family in the infection of Arabidopsis thaliana by potyviruses. *FEBS Lett.* **579**:1167–1171.

Schaad, M. C., Jensen, P. E., and Carrington, J. C. (1997). Formation of plant RNA virus replication complexes on membranes: role of an endoplasmic reticulum-targeted viral protein. *EMBO J.* **16**:4049–4059.

Schaad, M. C., Anderberg, R. J., and Carrington, J. C. (2000). Strain-specific interaction of the tobacco etch virus NIa protein with the translation initiation factor eIF4E in the yeast two-hybrid system. *Virology* **273**:300–306.

Schwartz, M., Chen, J., Janda, M., Sullivan, M., den Boon, J., and Ahlquist, P. (2002). A positive-strand rna virus replication complex parallels form and function of retrovirus capsids. *Mol. Cell* **9**:505–514.

Sharma, N., O'Donnell, B. J., and Flanegan, J. B. (2005). 3′-Terminal sequence in poliovirus negative-strand templates is the primary *cis*-acting element required for VPgpUpU-primed positive-strand initiation. *J. Virol.* **79**:3565–3577.

Slade, A., Fuerstenberg, S., Loeffler, D., Steine, M., and Facciotti, D. (2005). A reverse genetic, nontransgenic approach to wheat crop improvement by TILLING. *Nat. Biotechnol.* **23:**75–81.

Stein, N., Perovic, D., Kumlehn, J., Pellio, B., Stracke, S., Streng, S., Ordon, F., and Graner, A. (2005). The eukaryotic translation initiation factor 4E confers multiallelic recessive *Bymovirus* resistance in *Hordeum vulgare* (L.). *Plant J.* **42:**912–922.

Thivierge, K., Cotton, S., Dufresne, P. J., Mathieu, I., Beauchemin, C., Ide, C., Fortin, M. G., and Laliberté, J.-F. (2008). Eukaryotic elongation factor 1A interacts with *Turnip mosaic virus* RNA-dependent RNA polymerase and VPg-Pro in virus-induced vesicles. *Virology* **377:**216–225.

Truniger, V., Nieto, C., González-Ibeas, D., and Aranda, M. (2008). Mechanism of plant eIF4E-mediated resistance against a Carmovirus (*Tombusviridae*): cap-independent translation of a viral RNA controlled in cis by an (a)virulence determinant. *Plant J.* **56:**716–727.

Tsujimoto, Y., Numaga, T., Ohshima, K., Yano, M. a., Ohsawa, R., Goto, D., Naito, S., and Ishikawa, M. (2003). Arabidopsis TOBAMOVIRUS MULTIPLICATION (TOM) 2 locus encodes a transmembrane protein that interacts with TOM1. *EMBO J.* **22:**335–343.

Wang, X., Ullah, Z., and Grumet, R. (2000). Interaction between zucchini yellow mosaic potyvirus RNA-dependent RNA polymerase and host poly-(A) binding protein. *Virology* **275:**433–443.

Whitham, S. A., Yamamoto, M. L., and Carrington, J. C. (1999). Selectable viruses and altered susceptibility mutants in *Arabidopsis thaliana*. *Proc. Natl. Acad. Sci. U. S. A.* **96:**772–777.

Wittmann, S., Chatel, H., Fortin, M. G., and Laliberte, J.-F. (1997). Interaction of the viral protein genome linked of turnip mosaic potyvirus with the translational eukaryotic initiation factor (iso) 4E of *Arabidopsis thaliana* using the yeast two-hybrid system. *Virology* **234:**84–92.

Yamanaka, T., Ohta, T., Takahashi, M., Meshi, T., Schmidt, R., Dean, C., Naito, S., and Ishikawa, M. (2000). TOM1, an Arabidopsis gene required for efficient multiplication of a tobamovirus, encodes a putative transmembrane protein. *Proc. Natl. Acad. Sci. U. S. A.* **97:**10107–10112.

Yamanaka, T., Imai, T., Satoh, R., Kawashima, A., Takahashi, M., Tomita, K., Kubota, K., Meshi, T., Naito, S., and Ishikawa, M. (2002). Complete inhibition of tobamovirus multiplication by simultaneous mutations in two homologous host genes. *J. Virol.* **76:**2491–2497.

Yeam, I., Cavatorta, J. R., Ripoll, D. R., Kang, B.-C., and Jahn, M. M. (2007). Functional dissection of naturally occurring amino acid substitutions in eIF4E that confers recessive potyvirus resistance in plants. *Plant Cell* **19:**2913–2928.

Yi, G., Gopinath, K., and Kao, C. C. (2007). Selective repression of translation by the brome mosaic virus 1a RNA replication protein. *J. Virol.* **81:**1601–1609.

Yoshii, M., Yoshioka, N., Ishikawa, M., and Naito, S. (1998). Isolation of an Arabidopsis thaliana mutant in which accumulation of cucumber mosaic virus coat protein is delayed. *Plant J.* **13:**211–219.

Yoshii, M., Yoshioka, N., Ishikawa, M., and Naito, S. (1998). Isolation of an *Arabidopsis thaliana* mutant in which the multiplication of both cucumber mosaic virus and turnip crinkle virus is affected. *J. Virol.* **72:**8731–8737.
Yoshii, M., Nishikiori, M., Tomita, K., Yoshioka, N., Kozuka, R., Naito, S., and Ishikawa, M. (2004). The *Arabidopsis cucumovirus multiplication* 1 and 2 loci encode translation initiation factors 4E and 4G. *J. Virol.* **78:**6102–6111.
Yoshii, M., Shimizu, T., Yamazaki, M., Higashi, T., Miyao, A., Hirochika, H., and Omura, T. (2009). Disruption of a novel gene for a NAC-domain protein in rice confers resistance to *Rice dwarf virus*. *Plant J.* **57:**615–625.

CHAPTER 5

Toward a Quarter Century of Pathogen-Derived Resistance and Practical Approaches to Plant Virus Disease Control

J. Gottula and **M. Fuchs**

Contents

Department of Plant Pathology and Plant-Microbe Biology, Cornell University, New York State Agricultural Experiment Station, Geneva, NY 14456, USA

Advances in Virus Research, Volume 75
ISSN: 0065-3527, DOI: 10.1016/S0065-3527(09)07505-8

Abstract

The concept of pathogen-derived resistance (PDR) describes the use of genetic elements from a pathogen's own genome to confer resistance in an otherwise susceptible host via genetic engineering [*J. Theor. Biol.* 113 (1985) 395]. Illustrated with the bacteriophage Qβ in *Escherichia coli*, this strategy was conceived as a broadly applicable approach to engineer resistance against pathogens. For plant viruses, the concept of PDR was validated with the creation of tobacco plants expressing the coat protein gene of *Tobacco mosaic virus* (TMV) and exhibiting resistance to infection by TMV [*Science* 232 (1986) 738]. Subsequently, virus-resistant horticultural crops were developed through the expression of viral gene constructs. Among the numerous transgenic crops produced and evaluated in the field, papaya resistant to *Papaya ringspot virus* (PRSV) [*Annu. Rev. Phytopathol.* 36 (1998) 415] and summer squash resistant to *Cucumber mosaic virus* (CMV), *Zucchini yellow mosaic virus*, and/or *Watermelon mosaic virus* [*Biotechnology* 13 (1995) 1458] were released for commercial use in the USA. Although cultivated on limited areas, the adoption rate of cultivars derived from these two crops is increasing steadily. Tomato and sweet pepper resistant to CMV and papaya resistant to PRSV were also released in the People's Republic of China. Applying the concept of PDR provides unique opportunities for developing virus-resistant crops and implementing efficient and environmentally sound management approaches to mitigate the impact of virus diseases. Based on the tremendous progress made during the past quarter century, the prospects of further advancing this innovative technology for practical control of virus diseases are very promising.

I. INTRODUCTION

Plant viruses are responsible for severe economic crop losses worldwide (Hull, 2002). The development and use of resistant crop cultivars is the most efficient strategy to mitigate the impact of virus diseases in agricultural settings. Traditionally, host resistance is exploited by conventional breeding methods to create virus-resistant cultivars. Protection from virus infection can be achieved by using dominant or recessive genes. Examples of dominant resistance genes are *Ry* for *Potato virus Y* (PVY) in potato and *Sw5* for *Tomato spotted wilt virus* (TSWV) in tomato. The eukaryotic translation initiation factor (eIF4E) is an example of a recessive resistance gene for potyviruses (Kang *et al.*, 2005; Lanfermeijer and Hille, 2007).

The concept of pathogen-derived resistance (PDR) offers a different approach to develop virus-resistant crop plants. This concept was conceived a quarter century ago (Sanford and Johnston, 1985). It describes the engineering of resistance in otherwise susceptible hosts, including plants, by using genetic elements, for example coding and noncoding sequence elements, from a pathogen's own genome (Sanford and Johnston, 1985). Cross-protection, a biological means for protecting plants from virus infection, was considered as an example of the concept of PDR that is already operational in nature. Cross-protection relies on the use of mild virus strains to protect plants from economic damage caused by severe virus strains (Fuchs *et al.*, 1997; Muller and Rezende, 2004). It was argued that a mutated form of a viral replicase similar enough to the one encoded by a challenge virus could bind to cell host attachment sites and prevent virus replication (Sanford and Johnston, 1985).

For plant viruses, the concept of PDR was first validated with the development of tobacco expressing the coat protein gene of *Tobacco mosaic virus* (TMV) and exhibiting resistance to TMV infection (Powell Abel *et al.*, 1986). This breakthrough discovery paved the way for the creation of numerous virus-resistant transgenic plants, including horticultural crops. Some crop plants expressing viral genetic elements have been tested successfully in the field and a few have been commercialized. The deployment of virus-resistant transgenic plants has become an important strategy for effective and sustainable control of major virus diseases. This chapter provides a historical perspective on the concept of PDR from its inception to the release of the first virus-resistant transgenic crop resulting from its application. It also discusses how this concept led to an explosion in the development of virus-resistant plants and discusses advances made in terms of practical control of virus diseases during the past 25 years.

II. THE CONCEPT OF PDR

A. A description of the concept of PDR

The concept of PDR describes the use of a pathogen's own genetic material as resistance genes for engineering resistance in an otherwise susceptible host. Sanford and Johnston, the two visionary scientists who articulated the concept, initially proposed this strategy as a broadly applicable approach for genetically engineering resistance to parasites. Resistance was hypothesized to be routinely achievable by cloning appropriate parasite genes, modifying their expression, if necessary, and transferring them into the host genome.

Sanford and Johnston (1985) reasoned that pathogens produce molecules that are unique and critical for their pathogenic process. They proposed that dysfunctional pathogen-derived gene products could inhibit the pathogen by disrupting pathogen–host interactions if expressed by a host cell genome. To this extent, resistance could theoretically be achieved from the pathogen's own genetic material. The predicted advantages of this approach to engineer resistance in an otherwise susceptible host were that (i) genes from a pathogen would have a minimal effect on the host and likely not produce substances harmful to humans, (ii) the resistance was anticipated to be more stable than host resistance, (iii) cloning genes from a pathogen would be relatively easy compared to host genes due to the small genome size, and (iv) genes from a pathogen would be always present and available for cloning purposes regardless of the diversity of the pathogen. For plant viruses, the concept of PDR and successful transfer of foreign DNA into plant cells that regenerate into transgenic plants opened new avenues for the development of virus-resistant plants.

B. A conceivable application of the concept of PDR

Sanford and Johnston (1985) used the bacteriophage Qβ as a model to illustrate the concept of PDR. They hypothesized that four Qβ-encoded gene constructs, for example the coat protein gene, a modified replicase gene, the RNA segment encoding the replicase binding site, and the gene encoding the maturation protein, as well as an antisense RNA complementary to the Qβ RNA could be used as resistance genes against the bacteriophage Qβ in *Escherichia coli*. Sanford and Johnston (1985) further suggested that the strategy outlined for the bacteriophage Qβ in *Escherichia coli* could have a broader application for engineering resistance to other pathogens, opening an unsuspected path for practical control of diseases. To this extent, the stage was set for innovative ways to create virus-resistant plants by applying the concept of PDR and developing efficient protocols for plant transformation.

III. HISTORICAL PERSPECTIVES

A. The first application of the concept of PDR for virus resistance in a model host

Powell Abel *et al.* (1986) were the first to apply the concept of PDR to a plant virus. These authors produced *Nicotiana tabacum* cv. Xanthi and cv. Samsun expressing the coat protein gene of TMV and showed that transgenic tobacco exhibited resistance following infection by TMV via mechanical inoculation. Some transgenic tobacco failed to express symptoms for the duration of the experiments whereas others exhibited a substantial delay (2–14 days) in disease development. Resistance was related to the level of expression of the viral coat protein and could be overcome by high doses of inoculum under which conditions plants developed typical systemic symptoms and systemically infected leaves contained high TMV titer (Powell Abel *et al.*, 1986). Plants had only a slight enhanced resistance to TMV RNA as inoculum (Nelson *et al.*, 1987). The resistance was strong to tobamoviruses closely related to TMV but weak or not detectable to distantly related tobamoviruses (Nejidat and Beachy, 1990). Additional experiments suggested that increased levels of TMV coat protein expression correlated with increased levels of resistance (Osbourn *et al.*, 1989; Powell Abel *et al.*, 1989; Prins *et al.*, 2008; Register and Beachy, 1988, 1989). The initial intent of Beachy and colleagues for transferring and expressing the TMV coat protein gene into tobacco was to gain a better understanding of the mechanisms of cross-protection and provide new insights into virus–host interactions. Their seminal work launched a new era for the production of virus-resistant plants.

B. Other early applications of the concept of PDR for virus resistance

As a consequence of the discovery by Powell Abel *et al.* (1986), resistance to numerous plant viruses was engineered primarily by using coat protein genes (Beachy *et al.*, 1990; Prins *et al.*, 2008; Tepfer, 2002). Other viral sequences, such as the RNA-dependent RNA polymerase read-through domain of TMV were also shown to induce resistance (Golembowski *et al.*, 1990), as well as the movement protein (Malyshenko *et al.*, 1993), proteinase (Maiti *et al.*, 1993; Vardi *et al.*, 1993), satellite RNA (Gerlach *et al.*, 1987; Harrison *et al.*, 1987), defective interfering RNA (Kollar *et al.*, 1993), and 5′ (Nelson *et al.*, 1993; Stanley *et al.*, 1990) and 3′ (Zaccomer *et al.*, 1993) noncoding regions. It soon became apparent that almost any viral genetic element could be used to confer resistance to virus infection in plants. These observations validated some of the earlier predictions by Sanford and Johnston (1985) on the notion that several genes from a pathogen could be used to engineer resistance.

C. The concept of PDR and the antiviral pathways of RNA silencing

The mechanism of engineered resistance through the application of the concept of PDR was poorly understood 25 years ago. It was hypothesized that a dysfunctional viral gene in a host could somehow interfere with virus multiplication. By analogy with cross-protection, the mechanisms consisted conceivably of competition for host factors, inhibition of the uncoating of challenge virus (Sherwood, 1987), disruption of the replication of the challenge virus due to annealing of RNA species of the protective and challenge viruses (Palukaitis and Zaitlin, 1984), among other plausible explanations.

Expression of a viral coat protein in a transgenic plant was suggested initially to interfere with the uncoating step during an early event of the virus multiplication cycle (Osbourn *et al.*, 1989; Register and Beachy, 1988). Interaction of the viral coat protein with a host component or directly with the challenge viral RNA was hypothesized to prevent replication, translation, or virion assembly (Asurmendi *et al.*, 2007; Beachy, 1997, 1999; Bendahmane and Beachy, 1999; Clark *et al.*, 1995). Subsequently, a breakthrough discovery showed that an untranslatable coat protein gene of *Tobacco etch virus* (TEV) protected tobacco plants from TEV infection. Resistant plants were immune to TEV infection (Lindbo and Dougherty, 1992a, b). A recovery phenotype was also observed with plants infected and displaying symptoms similar to those of nontransgenic plants but newly emerging leaves were asymptomatic two weeks post-inoculation and transgene mRNA as well as viral RNA were rapidly degraded (Lindbo *et al.*, 1993). It became clear that the viral transgene protein product was not needed for engineered resistance and that there was an inverse correlation between transgene expression and resistance to virus infection (Dougherty *et al.*, 1994). In other words, the TEV coat protein RNA sequence was responsible for the resistance phenotype rather than the coat protein itself. This was unexpected as it was suggested that plants expressing high levels of viral coat protein would be likely resistant to virus infection in comparison with plants expressing little or no viral coat protein (Lindbo and Dougherty, 2005). Similar findings were published early on for TSWV (de Haan *et al.*, 1992) and *Potato virus X* (PVX) (Longstaff *et al.*, 1993) in tobacco plants. While coat protein-mediated resistance is effective against a number of viruses (Asurmendi *et al.*, 2007; Bendahmane and Beachy, 1999; Dinant *et al.*, 1998; Schubert *et al.*, 2004; Wintermantel and Zaitlin, 2000), the majority of PDR phenomena seem to work through RNA-mediated mechanisms (Baulcombe, 2007; Eamens *et al.*, 2008; Prins *et al.*, 2008; Voinnet, 2008).

Plant RNA-dependent RNA polymerase and double-stranded (ds) RNAase activities were proposed to be part of the mechanism of resistance by producing short RNA of 10–20 nt in length complementary in sequence to the RNA to be degraded from the transgene RNAs (Lindbo *et al.*, 1993). These short RNAs would target specific RNAs for degradation by a dsRNase activity (Dougherty and Parks, 1995). The studies by Dougherty and colleagues advanced our understanding of the mechanisms underlying engineered virus resistance in plants and highlighted the role of a sequence-specific RNA degradation phenomenon through post-transcriptional gene silencing (PTGS). Their findings paved the way to the discovery of RNA silencing as a potent defense mechanism against plant viruses (Baulcombe, 2004, 2007; Eamens *et al.*, 2008; Lin *et al.*, 2007; Prins *et al.*, 2008; Voinnet, 2001, 2005, 2008; Waterhouse *et al.*, 1999, 2001). Later, it was shown that antiviral silencing occurred during the recovery phase of virus infection in nontransgenic plants (Covey *et al.*, 1997; Ratcliff *et al.*, 1997).

RNA silencing is initiated by double-stranded RNA (dsRNA) structures that are identical to the RNA to be degraded (Waterhouse *et al.*, 1998). Silencing is associated with the production of 21–25 nt duplexes called small interfering RNAs (siRNAs) (Hamilton and Baulcombe, 1999; Hamilton *et al.*, 2002). The siRNAs are produced from dsRNA precursors by an endonuclease known as Dicer and become incorporated and converted to single-stranded RNAs (ssRNAs) in a Argonaute-containing ribonuclease complex (RISC) that target RNA for cleavage (Deleris *et al.*, 2006; Hannon, 2002; Obbard *et al.*, 2009; Voinnet, 2001, 2005, 2008). The pioneering work by Baulcombe and Waterhouse and their respective colleagues showed that RNA silencing is an innate and potent plant response to virus infection and a natural example of the concept of PDR.

D. The first application of the concept of PDR to a horticultural crop

Soon after its first application for virus resistance in a model host plant (Powell Abel *et al.*, 1986), the concept of PDR was validated in a horticultural crop with the aim of providing practical control of a viral disease. Tomato was the first horticultural crop engineered for virus resistance through the application of the concept of PDR. In the first field trial ever of transgenic plants engineered for virus resistance, tomato plants expressing the coat protein gene of TMV were evaluated for resistance to mechanical inoculation by TMV (Nelson *et al.*, 1988). Only 5% of the transgenic plants were symptomatic at the end of the trial compared with 99% of the nontransformed control plants. Also, inoculated transgenic and uninoculated nontransformed plants had identical fruit yield, indicating that the transformation process and expression of the TMV coat protein gene did not alter the horticultural performance of the transgenic tomato plants. Sanders *et al.* (1992) extended the field characterization of transgenic tomato plants and showed resistance to distinct strains of TMV. These studies confirmed Sanford and Johnston's conception of PDR as a practical solution for controlling virus diseases in plants.

IV. CREATION OF VIRUS-RESISTANT TRANSGENIC CROPS BY APPLYING THE CONCEPT OF PDR

A. Early applications

Effective resistance is desirable against virus inoculation via vectors to manage, for instance, aphid-transmitted virus diseases. The efficiency of viral genes at conferring resistance against vector-mediated virus transmission was shown first with cucumber plants engineered for resistance to *Cucumber mosaic virus* (CMV). Cucumber plants expressing the coat protein gene of CMV had a significantly reduced incidence of CMV and a lower percentage of symptomatic plants than nontransformed control plants following CMV inoculation via aphid vectors (Gonsalves *et al.*, 1992). In these studies, mechanically inoculated cucumber plants dispersed throughout the field provided reliable sources of inoculum for natural aphid populations to vector CMV. This approach coupled with the fact that field trials were established at a time of abundant endemic aphid flights caused sufficient disease pressure to make inferences about disease progress, resistance, and yield (Gonsalves *et al.*, 1992). Subsequently, many other studies have illustrated the usefulness of engineered resistance at providing practical control of aphid-transmitted virus diseases (reviewed by Fuchs and Gonsalves, 2007).

Resistance to more than one virus is useful for practical control of virus diseases as mixed virus infections are common in agricultural settings. PDR offers unique solutions to mixed virus infection, for example, by co-engineering and co-transferring genes from several viruses into a single host plant. The usefulness of multiple viral genes to control mixed virus infections was demonstrated early on with potato plants expressing the coat protein genes of PVX and PVY (Kaniewski *et al.*, 1990; Lawson *et al.*, 1990). Potato line 303 was highly resistant to infections by PVX and PVY in the field (Kaniewski *et al.*, 1990). Later, summer squash plants expressing coat protein gene constructs of CMV, *Zucchini yellow mosaic virus* (ZYMV), and/or *Watermelon mosaic virus* (WMV) were engineered for resistance to single viruses and combinations of these three viruses (Tricoli *et al.*, 1995). Among summer squash engineered for multiple virus resistance, line ZW-20 expressing the coat protein genes of ZYMV, and WMV was highly resistant whether infection occurred by mechanical inoculation or was mediated by aphid vectors (Fuchs and Gonsalves, 1995; Tricoli *et al.*, 1995). In addition, line CZW-3 expressing the coat protein genes of CMV, ZYMV and WMV was highly resistance to mixed infections by these three viruses (Fuchs *et al.*, 1998; Tricoli *et al.*, 1995). The three coat protein genes used to engineer multiple virus resistance in summer squash were transferred successfully in cantaloupes (Fuchs *et al.*, 1997). The concept of PDR has provided a platform for virus control that has facilitated new approaches to develop resistant crop cultivars and expanded opportunities to implement effective and sustainable management strategies of virus diseases.

B. Other examples

Agronomic and horticultural plants, such as cereal, vegetable, legume, flower, forage, turf, and fruit crops expressing virus-derived gene constructs have been created (Fuchs and Gonsalves, 2007). While testing in the field is underway for at least one dozen crop species expressing sequences derived from numerous viruses, very few of these field trials have yet been published in scientific journals.

Part of the difficulty of field-testing for resistance evaluation is ensuring high and consistent virus inoculation that can distinguish resistant and susceptible phenotypes (Gilbert *et al.*, 2009). Inoculation from external field sources can be reliable when studies are with insect-transmitted viruses. Presence of naturally viruliferous aphid populations allowed Lee *et al.* (2009) to discern resistant pepper expressing a CMV coat protein gene in conditions relevant to commercial agriculture. Natural infection by thrips vectors yielded statistically significant differences in TSWV incidence in peanut expressing an antisense TSWV nucleoprotein sequence (Magbanua *et al.*, 2000). In another study, peanut expressing a TSWV nucleoprotein had a strong tendency to be asymptomatic under field locations, although resistance was moderate following mechanical inoculation in a growth chamber (Yang *et al.*, 2004). Transgenic peanut and pepper showed good yield and quality parameters, respectively (Lee *et al.*, 2009; Yang *et al.*, 2004). For perennial crops, plum trees expressing a coat protein gene construct of *Plum pox virus* (PPV) were highly resistant to PPV infection during 6–8 years in varied orchard locations in Europe (Capote *et al.*, 2007; Hily *et al.*, 2004; Malinowski *et al.*, 2006; Ravelonandro, 2007; Zagrai *et al.*, 2008). The growth of knowledge about RNA silencing has provided a basis to optimize constructs to make engineered resistance to viruses more reliable and broadly applicable.

C. More recent applications

The trigger for RNA silencing is dsRNA or double-stranded regions within the secondary structure of single-stranded RNA (Eamens *et al.*, 2008; Prins *et al.*, 2008; Voinnet, 2008; Waterhouse *et al.*, 1998). Several approaches have been used to express dsRNA cognate to viral RNA for activation of RNA silencing. Expressing sense and antisense viral genes or inverted repeat viral genes to express hairpin RNAs (hpRNA) for the formation of duplex RNA are some of the most recent strategies to engineer resistance (Missiou *et al.*, 2004; Praveen *et al.*, 2009; Prins *et al.*, 2008; Smith *et al.*, 2000; Tougou *et al.*, 2006; Wesley *et al.*, 2001). For example, intron-spliced hairpin RNA (ihpRNA), ihpRNA overhang, and ihpRNA spacer were evaluated for resistance to PVY (Smith *et al.*, 2000; Wesley *et al.*, 2001). The ihpRNA was found to be the most efficient constructs to conferring resistance to PVY with 90% of the plants exhibiting RNA silencing (Wesley *et al.*, 2001). The same strategy based on the use of highly conserved genetic segments of several viruses into a single transgene construct achieved multiple virus resistance (Bucher *et al.*, 1996).

Artificial plant micro RNA (amiRNAs) can also be used for virus resistance. The *Arabidopsis thaliana* pre-miR159a precursor was used to generate two amiRNAs^{159} (amiR-P69^{159} and amiR-HC-Pro^{159}) with sequences complementary to *Turnip yellow mosaic virus* (TYMV) and *Turnip mosaic virus* (TuMV), respectively (Niu *et al.*, 2006). The amiR-P69159 was designed to target the TYMV silencing suppressor P69 while amiR-HC-Pro^{159} targeted the TuMV silencing suppressor HC-Pro. Transgenic plants carrying both transgenes expressed the corresponding amiRNAs and showed specific resistance to TYMV and TuMV. Low temperatures had no substantial effect on miRNA accumulation (Niu *et al.*, 2006). Similarly, the miR171 of *Nicotiana benthamiana* was used to target the 2b gene of CMV and confer resistance to CMV (Qu *et al.*, 2008).

V. COMMERCIALIZATION OF VIRUS-RESISTANT TRANSGENIC CROPS AND PRACTICAL CONTROL OF VIRUS DISEASES

A. Virus-resistant summer squash

Summer squash expressing the CP gene of ZYMV and WMV received exemption status in the USA in 1994 and was released thereafter. This was the first disease-resistant transgenic crop to be commercialized in the USA (Table 1). Plants of line ZW-20 are vigorous following exposure to aphid-mediated transmission of ZYMV and WMV (Fig. 1a) and produce marketable fruits (Fig. 1b) unlike conventional squash. Summer squash expressing the CP gene of CMV, WMV, and ZYMV was deregulated and commercialized in 1996. Subsequently, numerous squash types and cultivars have been developed by crosses and backcrosses with the two initially deregulated lines. Currently there are five zucchini and six straightneck or crookneck yellow squash cultivars for which combinations of resistance to ZYMV and WMV or resistance to CMV, ZYMV, and WMV are available.

TABLE 1 Successful application of PDR in commercially available virus-resistant crops*

Crop	Scientific name	Resistance to	Country of release
Papaya	*Carica papaya*	*Papaya ringspot virus*	USA, People's Republic of China
Pepper	*Capsicum*	*Cucumber mosaic virus*	People's Republic of China
Squash	*Cucurbita pepo*	*Cucumber mosaic virus* *Watermelon mosaic virus* *Zucchini yellow mosaic virus*	USA
Tomato	*Solanum lycopersicum*	*Cucumber mosaic virus*	People's Republic of China

*From James (2009) and Stone (2008).

FIGURE 1 Reaction of summer squash and papaya to virus infection. (a) Resistance of transgenic summer squash ZW-20 (center and right rows) to aphid-mediated transmission of ZYMV and WMV from virus-infected conventional plants that served as inoculum source following mechanical inoculation (left row and first plant in the center row). (b) Comparative fruit yield of virus-resistant transgenic summer squash (back) and virus-infected conventional squash (front). (c) Aerial view of an experimental field of healthy transgenic PRSV-resistant Rainbow papaya (center) surrounded by rows of PRSV-infected conventional papaya (courtesy of D. Gonsalves). (d) Commercial field of PRSV-resistant papaya field in Hawaii.

The adoption of virus-resistant summer squash cultivars is steadily increasing since 1996. In 2006, the adoption rate was estimated to 22% (3,250 hectares) across the country with an average rate of 70% in New Jersey and 20% in Florida, Georgia, and South Carolina. The benefit to growers was estimated to $24 million in 2006 (Johnson *et al.*, 2007).

B. Virus-resistant papaya

Papaya expressing the coat protein gene of *Papaya ringspot virus* (PRSV) was deregulated in 1998 and commercialized in Hawaii (Table 1). PRSV is a major limiting factor to papaya production in Hawaii and around the world. After extensive experimental testing (Fig. 1c), PRSV-resistant papaya was released in 1998 as devastation caused by the virus reached record proportions in the archipelago's main production region (Gonsalves, 1998). The impact of PRSV-resistant papaya on the papaya industry in Hawaii is evidenced by its rapid adoption rate (Fig. 1d). In 2000, the first wave of transgenic papaya bore fruit on more than 42% of the total acreage (Johnson *et al.*, 2007). Resumption of fruitful harvests put papaya packing houses back in business and provided a $4.3 million impact over a 6-year period (Fuchs, 2008).

By 2006, transgenic papaya cultivars were planted on more than 90% of the total papaya land in Hawaii (780 of 866 total hectares) (Johnson *et al.*, 2007), with the remaining conventional fruit shipped mainly to Japan, one of the major export countries for the Hawaiian papaya industry along with Canada (Suzuki *et al.*, 2007). After a decade of segregating transgenic and nontransgenic papaya fruits, this practice may be nearing end due to the recent deregulation of the transgenic fruit in Japan (D. Gonsalves, personal communication) following deregulation in Canada (Suzuki *et al.*, 2007).

C. Other examples

Two virus-resistant potato lines were deregulated in 1998 and 2000 in the USA. After failed attempts to create a potato line resistant to *Potato leafroll virus* (PLRV) by coat protein gene expression, lines expressing a PLRV replicase gene were created, field tested, deregulated, and commercialized (Kaniewski and Thomas, 2004). Later, this resistance was stacked with a synthetic Bt gene that conferred resistance to Colorado potato beetle. Another potato cultivar was developed by adding the coat protein gene of PVY. Although many growers in the Pacific Northwest, Midwest US and Canada were growing transgenic potato, and no resistance breakage was reported, nor any detrimental impact on the environmental or human health, virus-resistant potato were withdrawn from the market after the 2001 season due to the reluctance of several large processors and exporters to adopt these products (Kaniewski and Thomas, 2004).

In the People's Republic of China, tomato and sweet pepper resistant to CMV were released as well as papaya resistant to PRSV (James, 2009; Stone, 2008) (Table 1). Limited if any, information is available on their adoption rate.

Although not released yet, the plum cultivar 'Honeysweet' resistant to PPV is under consideration for deregulation in the USA. The US Department of Agriculture (USDA) Animal and Plant Health Inspection Service (APHIS) has granted this cultivar deregulated status (Bech, 2007) and the Food and Drug Administration (FDA) has deemed a pre-market review of the 'Honeysweet' unnecessary. Presently, the Environmental Protection Agency (EPA) is examining deregulation petitions for 'Honeysweet.' Another PRSV-resistant papaya has been deregulated by two of the three US biotechnology regulatory authorities. Line X17-2 differs from the previously deregulated Hawaiian papaya in that it expresses the CP gene of a Florida isolate of PRSV and is suitable for cultivation in Florida (Davis, 2004). APHIS and the FDA have granted X17-2 deregulated status (Anonymous, 2009; Shea, 2009). The realized economic benefits and minimal environmental hazards of the previously deregulated virus-resistant Hawaiian papaya figured prominently into APHIS' favorable consideration (Gregoire and Abel, 2008). The EPA will consider the plant pest risk of X17-2 after the developer submits a petition for deregulation.

D. Stability and durability of engineered virus resistance

Plant viruses can evade the antiviral defense response by encoding RNA silencing-suppressor genes (Díaz-Pendón and Ding, 2008; Ding and Voinnet, 2007; Eamens *et al.*, 2008; Li and Ding, 2006; Voinnet, 2008). The HC-Pro protein of TEV and the 2b protein of CMV were amongst the first viral suppressors of transgene silencing identified (Anandalakshmi *et al.*, 1998; Brigneti *et al.*, 1998; Kasschau and Carrington, 1998). Silencing suppressors from different plant viruses counteract various steps in the RNA silencing process (Díaz-Pendón and Ding, 2008; Ding and Voinnet, 2007; Li and Ding, 2006; Voinnet *et al.*, 1999). As a consequence, silencing-based resistance to one virus can be partially counteracted by infection with an unrelated virus carrying a silencing-suppressor gene (Mitter *et al.*, 2003). Such an effect was not observed with plum trees expressing the PPV CP gene (Ravelonandro, 2007; Zagrai *et al.*, 2008) or summer squash expressing the CP genes of CMV, ZYMV, and WMV (Fuchs *et al.*, 1998; Tricoli *et al.*, 1995) following infection with heterologous viruses.

Transcriptional gene silencing and genetic background are two documented variables that can cause transgenic plants expressing viral sequences to lack the expected resistance phenotype (Febres *et al.*, 2008). Wheat transformed with the coat protein gene or replicase gene of *Wheat streak mosaic virus* tended to display more severe symptoms and higher relative virus titers in the field compared to the nontransformed parent cultivar (Sharp *et al.*, 2002). These results suggested that environmental conditions can affect the stability of engineered resistance since all transgenic lines showed a recovery phenotype in greenhouse experiments. It is known that RNA silencing is inactive at low temperatures (Szittya *et al.*, 2003). This provides good conceptual rationale to pyramid virus-derived transgenes with conventional resistance genes, which can be inactivated at high temperature (Wang *et al.*, 2009).

Resistance breakage has not been reported in more than 10 years of commercial deployment of transgenic summer squash and papaya. Similarly, resistance is durable for PPV-resistant plum trees tested in experimental orchards over 13 years in Europe, despite constant exposure to viruliferous aphids vectoring diverse PPV populations (Capote *et al.*, 2007; Malinowski *et al.*, 2006; Ravelonandro, 2007; Zagrai *et al.*, 2007). In contrast, resistance breakdown has been demonstrated in laboratory and greenhouse settings with papaya. Early work revealed that resistance to PRSV was narrow in cultivars expressing the CP gene from a Hawaiian isolate of PRSV; plants were resistant to PRSV isolates from Hawaii but largely susceptible to isolates outside of Hawaii, depending on the extent of sequence divergence (Suzuki *et al.*, 2007). Efforts to pyramid genes from highly conserved region of the PRSV genome from various isolates for broad-spectrum resistance are under way.

VI. DISCUSSION

The concept of PDR was described a quarter century ago (Sanford and Johnston, 1985). This theory has provided a framework to engineer genetic constructs from a viral genome and use them as resistance genes to protect plants from virus infection. Application of this conceptual knowledge has introduced novel approaches for virus control by providing new means to develop resistant crop cultivars and increase opportunities to implement effective and sustainable management strategies of virus diseases. After its validation with TMV in tobacco plants in 1986, the concept of PDR has been applied successfully against a wide range of viruses in many plant species so that the past 25 years have witnessed an explosion in the development of virus-resistant transgenic plants.

Several virus-resistant transgenic crops resulting from the application of PDR have been extensively evaluated under field conditions and many more have been created and validated in laboratory or greenhouse conditions. The first resistant horticultural crops resulting from the application of PDR were vegetable (summer squash, sweet pepper, tomato, and potato) and fruit (papaya and plum) crops. Based on their efficacy at controlling virus diseases (Eamens *et al.*, 2008; Hily *et al.*, 2004; Prins *et al.*, 2008; Suzuki *et al.*, 2007; Tricoli *et al.*, 1995), a history of ready adoption by growers (Suzuki *et al.*, 2007) and no documented detrimental environmental impact (Fuchs and Gonsalves, 2007), more virus-resistant transgenic crops are likely to reach the market in the future. While several crop plants show good resistance to virus infection in the field, the dearth of commercialized examples beyond summer squash, papaya, tomato, and sweet pepper suggests that steep legal or regulatory issues, among other issues, have barred market entry.

The dedication and perseverance on the part of a handful of researchers in the public and private sectors have extended PDR beyond an academic exercise to a proven technology for commercial use and efficient management of virus diseases. The creation and deployment of PRSV-resistant papaya have provided a safe and effective way to save an entire fruit industry on the Hawaiian Islands. The same could be true for Thailand but for negative intervention by an international nongovernmental organization (Davidson, 2008). Virus-resistant summer squash and potato have been deregulated in the USA but only summer squash remain commercially available to date. A virus-resistant plum and another virus-resistant papaya await full deregulation in the USA. The People's Republic of China is likewise moving forward with virus-resistant transgenic crops and has already commercialized virus-resistant sweet pepper, tomato, and papaya (James, 2009; Stone, 2008).

The application of the concept of PDR also paved the way for tremendous progress to be made at unraveling the biology of antiviral pathways of RNA silencing in plants, a natural and potent defense mechanism against viruses that can be triggered by the insertion and expression of viral gene constructs in susceptible hosts (Baulcombe, 2007; Eamens *et al.*, 2008; Lin *et al.*, 2007; Obbard *et al.*, 2009; Prins *et al.*, 2008; Voinnet, 2008). Knowledge of RNA silencing has provided new and unprecedented insights into virus–host interactions. dsRNA was identified as trigger of the antiviral defense mechanism, virus-encoded silencing suppressors as counterattack factors and symptom inducers, and pathogen-homing siRNAs as guides for the destruction of viral RNA by RISC (Baulcombe, 2007; Eamens *et al.*, 2008; Lin *et al.*, 2007; Obbard *et al.*, 2009; Voinnet, 2008). These developments stemming from the theory of PDR (Sanford and Johnston, 1985) shed light on the molecular and cellular mechanisms underlying engineered resistance in plants expressing virus-derived gene constructs.

The concept of PDR (Sanford and Johnston, 1985) provided unique opportunities for innovative solutions to control virus diseases by developing virus-resistant crops expressing genetic elements derived from a virus' own genome. A quarter century later, lessons from field experiments with various transgenic crops engineered for virus resistance and the commercial release of virus-resistant papaya, summer squash, sweet pepper, and tomato have conclusively demonstrated that applying the concept PDR is a practical strategy to mitigate the impact of virus diseases on agriculture.

ACKNOWLEDGMENTS

We are grateful to Dr. John Carr and Dr. Gad Loebenstein for helpful comments and to Joe Ogrodnick for artwork.

REFERENCES

Anandalakshmi, R., Pruss, G. J., Ge, X., Marathe, R., Mallory, A. C., Smith, T. H., and Vance, V. B. (1998). A viral suppressor of gene silencing in plants. *Proc. Natl. Acad. Sci. U. S. A.* **95:**13079–13084.

Anonymous. (2009). *Completed consultations on bioengineered foods*. U.S. Food and Drug Administration. [http://www.fda.gov/Food/Biotechnology/Submissions/default.htm].

Asurmendi, S., Berg, R. H., Smith, T. J., Bendahmane, M., and Beachy, R. N. (2007). Aggregation of TMV CP plays a role in CP functions and in coat-protein-mediated resistance. *Virology* **366:**98–106.

Bech, R. (2007). Finding of no significant impact and decision notice. *Determination of nonregulated status for C5 plum resistant to plum pox virus*. Animal and Plant Health Inspection Service. [http://www.aphis.usda.gov/brs/aphisdocs2/04_26401p_com.pdf].

Baulcombe, D. (2004). RNA silencing in plants. *Nature* **431:**356–363.

Baulcombe, D. (2007). Amplified silencing. *Science* **315:**199–200.

Beachy, R. N. (1997). Mechanisms and application of pathogen-derived resistance in transgenic plants. *Curr. Opin. Plant Biotechnol.* **8:**215–220.

Beachy, R. N. (1999). Coat protein-mediated resistance to tobacco mosaic virus: Discovery mechanisms and exploitation. *Phil. Trans. R. Soc. Lond. B.* **354:**659–664.

Beachy, R. N., Loesch-Fries, S., and Tumer, N. E. (1990). Coat protein-mediated resistance against virus infection. *Annu. Rev. Phytopathol.* **28:**451–474.

Bendahmane, M., and Beachy, R. N. (1999). Control of tobamovirus infections via pathogen-derived resistance. *Adv. Virus Res.* **53:**369–386.

Brigneti, G., Voinnet, O., Li, W., Ji, L., Ding, S. W., and Baulcombe, D. (1998). Viral pathogenicity determinants are suppressors of transgene silencing in *Nicotiana benthamiana*. *EMBO J.* **17:**6739–6746.

Bucher, E., Lohuis, D., Van Poppel, P. M. J. A., Geerts-Dimitriadou, C., Golbach, R., and Prins, M. (1996). Multiple virus resistance at a high frequency using a single transgene construct. *J. Gen. Virol.* **87:**3697–3701.

Capote, N., Pérez-Panadés, J., Monzó, C., Carbonell, E., Urbaneja, A., Scorza, R., Ravelonandro, M., and Cambra, M. (2007). Assessment of the diversity and dynamics of *Plum pox virus* and aphid populations in transgenic European plums under Mediterranean conditions. *Trans. Res.* **17**:367–377.

Clark, W. G., Fitchen, J. H., and Beachy, R. N. (1995). Studies of coat-protein-mediated resistance to TMV. I. The PM2 assembly defective mutant confers resistance to TMV. *Virology* **208**:485–491.

Covey, S. N., Al-Kaff, N. S., Langara, A., and Turner, D. S. (1997). Plants combat infection by gene silencing. *Nature* **385**:781–782.

Davidson, S. N. (2008). Forbidden fruit: Transgenic papaya in Thailand. *Plant Physiol.* **147**:487–493.

Davis, M.J. (2004). Petition for Determination of Nonregulated status for the X17-2 line of papaya: A *Papaya ringspot virus*-resistant papaya. Animal and Plant Health Inspection Service. [http://www.aphis.usda.gov/brs/aphisdocs/04_33701p.pdf].

de Haan, P., Gielen, J. J. L., Prins, M., Wijkamp, I. G., van Shepen, A., Peters, D., van Grinsven, M. Q. J. M., and Goldbach, R. (1992). Characterization of RNA-mediated resistance to Tomato spotted wilt virus in transgenic tobacco plants. *Biotechnology* **10**:1133–1137.

Deleris, A., Gallego-Bartolome, J., Bao, J., Kasschau, K. D., Carrington, J. C., and Voinnet, O. (2006). Hierarchical action and inhibition of plant dicer-like proteins in antiviral defense. *Science* **313**:68–71.

Díaz-Pendón, J. A., and Ding, S-W. (2008). Direct and indirect roles of viral suppressors of RNA silencing in pathogenesis. *Annu. Rev. Phytopathol.* **46**:303–326.

Dinant, S., Kusiak, C., Cailleteau, B., Verrier, J. L., Chupeau, M. C., Chupeau, Y., Le, Thi Ahn Hong, Delon, R., and Albouy, J. (1998). Field resistance against potato virus Y infection using natural and genetically engineered resistance genes. *Eur. J. Plant Pathol.* **104**:377–382.

Ding, S. W., and Voinnet, O. (2007). Antiviral immunity directed by small RNAs. *Cell* **130**:413–426.

Dougherty, W. G., and Parks, T. D. (1995). Transgenes and gene suppression: Telling us something new? *Curr. Opin. Cell Biol.* **7**:399–405.

Dougherty, W. G., Lindbo, J. A., Smith, H. A., Parks, T. F., Swaney, S., and Proebsting, W. M. (1994). RNA-mediated virus resistance in transgenic plants: Exploitation of a cellular pathway possibly involved in RNA degradation. *Mol. Plant-Microbe Interact.* **7**:544–552.

Eamens, A., Wang, M-B., Smith, N. A., and Waterhouse, P. M. (2008). RNA silencing in plants: Yesterday, today and tomorrow. *Plant Physiol.* **147**:456–468.

Febres, V. J., Lee, R. F., and Moore, G. A. (2008). Transgenic resistance to *Citrus tristeza virus* in grapefruit. *Plant Cell Rep.* **27**:93–104.

Fuchs, M. (2008). Plant resistance to viruses: Engineered resistance. *In* "Encyclopedia of Virology" (B. W. J. Mahy and M. van Regenmortel, eds.), pp. 156–164. Academic Press, San Diego, CA.

Fuchs, M., and Gonsalves, D. (1995). Resistance of transgenic squash Pavo ZW-20 expressing the coat protein genes of zucchini yellow mosaic virus and watermelon mosaic virus 2 to mixed infections by both potyviruses. *Bio/Biotechnol.* **13**:1466–1473.

Fuchs, M., and Gonsalves, D. (2007). Safety of virus-resistant transgenic plants two decades after their introduction: Lessons from realistic field risk assessment studies. *Annu. Rev. Phytopathol.* **45:**173–202.

Fuchs, M., Ferreira, S., and Gonsalves, D. (1997). Management of virus diseases by classical and engineered protection. *Mol. Plant Pathol. On-line* [http://www.bspp.org.uk/mppol/1997/0116fuchs/]

Fuchs, M., Tricoli, D. M., McMaster, J. R., Carney, K. J., Schesser, M., McFerson, J. R., and Gonsalves, D. (1998). Comparative virus resistance and fruit yield of transgenic squash with single and multiple coat protein genes. *Plant Dis.* **82:**1350–1356.

Gerlach, W. L., Llewellyn, D., and Haseloff, J. (1987). Construction of a disease resistant gene using the satellite RNA to tobacco ringspot virus. *Nature* **328:**802–805.

Gibert, R. A., Glynn, N. C., Comstock, J. C., and Davis, M. J. (2009). Agronomic performance and genetic characterization of sugarcane transformed for resistance to *Sugarcane yellow leaf virus*. *Field Crops Res.* **111:**39–46.

Golembowski, D. B., Lomonossoff, G. P., and Zaitlin, M (1990). Plants transformed with tobacco mosaic virus nonstructural gene sequence are resistant to the virus. *Proc. Natl. Acad. Sci. U. S. A.* **87:**6311–6315.

Gonsalves, D. (1998). Control of *Papaya ringspot virus* in papaya: A case study. *Annu. Rev. Phytopathol.* **36:**415–437.

Gonsalves, D., Chee, P., Provvidenti, R., Seem, R., and Slightom, J. L. (1992). Comparison of coat protein-mediated and genetically-derived resistance in cucumbers to infection by *Cucumber mosaic virus* under field conditions with natural challenge inoculations by vectors. *Biotechnology* **10:**1562–1570.

Gregoire, M., and Abel, S. (2008). USDA/APHIS draft environmental assessment. In response to University of Florida petition 04-337-01P seeking a determination of nonregulated status for X17-2 papaya resistant to *Papaya ringspot virus*. Animal and Plant Health Inspection Service. [http://www.aphis.usda.gov/brs/aphisdocs/04_33701p_pea.pdf].

Hamilton, A. J., and Baulcombe, D. C. (1999). A species of small antisense RNA in post-transcriptional gene silencing in plants. *Science* **286:**950–952.

Hamilton, A. J., Voinnet, O., Chappell, L., and Baulcombe, D. C. (2002). Two classes of short interfering RNA in RNA silencing. *EMBO J.* **21:**4671–4679.

Hannon, G. J. (2002). RNA interference. *Nature* **418:**244–251.

Harrison, B. D., Mayo, M. A., and Baulcombe, D. C. (1987). Virus resistance in transgenic plants that express *Cucumber mosaic virus* satellite RNA. *Nature* **328:**799–802.

Hily, J. M., Scorza, R., Malinowski, T., Zawadzka, B., and Ravelonandro, M. (2004). Stability of gene silencing-based resistance to *Plum pox virus* in transgenic plum (*Prunus domestica* L.) under field conditions. *Trans. Res.* **5:**427–436.

Hull, R. (2002). Economic Losses due to Plant Viruses. *In* "Matthew's Plant Virology" (R. Hull, ed.), Academic Press, New York, NY, USA.

James, C. (2009). Global status of commercialized biotech/GM crops: 2008. *The first thirteen years, 1996 to 2008.* International Service for the Acquisition of Agri-Biotech Applications. [http://www.isaaa.org/resources/publications/briefs/39/executivesummary/default.html]

Johnson, S.R., Strom, S., and Grillo, K. (2007). Quantification of the impacts on US agriculture of biotechnology-derived crops planted in 2006. National Center for Food and Agriculture policy. http://www.ncfap.org.

Kang, B. C., Yeam, I., and Jahn, M. M. (2005). Genetics of plant virus resistance. *Annu. Rev. Phytopathol.* **43:**581–621.

Kaniewski, W. K., and Thomas, P. E. (2004). The potato story. *Agric. Biol. Forum* **7:**41–46.

Kaniewski, W., Lawson, C., Sammons, B., Haley, L., Hart, J., Delannay, X., and Tumer, N. (1990). Field resistance of transgenic Russet Burbank potato to effects of infection by *Potato virus X* and *Potato virus Y. Biotechnology* **8:**750–754.

Kasschau, K. D., and Carrington, J. C. (1998). A counterdefensive strategy of plant viruses: Suppression of posttranscriptional gene silencing. *Cell* **95:**461–470.

Kollar, A., Dalmay, T., and Burgyan, J. (1993). Defective interfering RNA mediated resistance against cymbidium ringspot tombusvirus in transgenic plants. *Virology* **193:**313–318.

Lanfermeijer, F. C. and Hille, J. (2007). Molecular characterization of endogenous plant virus resistance genes. *In* "Biotechnology and Plant Disease Management" (Z. K. Punja, S. H. De Boer and H. Sanfaçon, eds.), pp. 395–415.

Lawson, C., Kaniewski, W., Haley, L., Rozman, R., Newell, C., Sanders, P., and Tumer, N. (1990). Engineering resistance to mixed virus infection in a commercial potato cultivar: Resistance of potato virus X and potato virus Y in transgenic Russet Burban. *Biotechnology* **8:**127–134.

Lee, Y. H., Jung, M., Shin, S. H., Lee, J. H., Choi, S. H., Her, N. H., Lee, J. H., Ryu, K. H., Paek, K. Y., and Harn, C. H. (2009). Transgenic peppers that are highly tolerant to a new CMV pathotype. *Plant Cell Rep.* **28:**223–232.

Li, F., and Ding, S-W. (2006). Virus counterdefense: Diverse strategies for evading the RNA-silencing immunity. *Annu. Rev. Microbiol.* **60:**503–531.

Lin, S-S., Henriques, R., Wu, H-W., Niu, Q-W., Yeh, S-D., and Chua, N-H. (2007). Strategies and mechanisms of plant virus resistance. *Plant Biotechnol. Rep.* **1:**125–134.

Lindbo, J. A., and Dougherty, W. G. (1992). Untranslatable transcripts of the tobacco etch virus coat protein gene sequence can interfere with tobacco etch virus replication in transgenic plants and protoplasts. *Virology* **189:**725–733.

Lindbo, J. A., and Dougherty, W. G. (1992). Pathogen-derived resistance to a potyvirus: Immune and resistant phenotypes in transgenic tobacco expressing altered form of a potyvirus coat protein nucleotide sequence. *Mol. Plant-Microbe Interact.* **5:**144–153.

Lindbo, J. A., and Dougherty, W. G. (2005). Plant Pathology and RNAi: A brief history. *Annu. Rev. Phytopathol.* **43:**191–204.

Lindbo, J. A., Silva-Rosales, L., Proebsting, W. M., and Dougherty, W. G. (1993). Induction of a highly specific antiviral state in transgenic plants: Implications for regulation of gene expression and virus resistance. *Plant Cell* **5:**1749–1759.

Longstaff, M., Brigneti, G., Bocard, F., Chapman, S., and Baulcombe, D. (1993). Extreme resistance to potato virus X infection in plants expressing a modified component of the putative viral replicase. *The EMBO J.* **12:**379–386.

Magbanua, Z. V., Wilde, H. D., Roberts, J. K., Chowdhury, K., Abad, J., Moyer, J., Wetzstein, H. Y., and Parrott, W. A. (2000). Field resistance to *Tomato spotted wilt virus* in transgenic peanut (*Arachis hypogaea* L.) expressing an antisense nucleocapsid gene sequence. *Mol. Breeding* **6:**227–236.

Maiti, I. B., Murphy, J. F., Shaw, J. G., and Hunt, A. G. (1993). Plants that express a potyvirus proteinase gene are resistant to virus infection. *Proc. Natl. Acad. Sci. U. S. A.* **90:**6110–6114.

Malinowski, T., Cambra, M., Capote, N., Zawadzka, B., Gorris, M. T., and Ravelonandro, M. (2006). Field trials of plum clones transformed with the Plum pox virus coat protein (PPV-CP) gene. *Plant Dis.* **90:**1012–1018.

Malyshenko, S. I., Kondakova, O. A., Nazarova, J. V., Kaplan, I. B., Taliansky, M. E., and Atabekov, J. G. (1993). Reduction of tobacco mosaic virus accumulation in transgenic plants producing non dysfunctional viral transport proteins. *J. Gen. Virol.* **74:**1149–1156.

Mitter, N., Sulistyowati, E., and Dietzgen, R. G. (2003). Cucumber mosaic virus infection transiently breaks dsRNA-induced transgenic immunity to *Potato virus Y* in tobacco. *Mol. Plant-Microbe Interact.* **16:**936–944.

Missiou, A., Kalantidis, K., Boutla, A., Tzortzakaki, S., Tabler, M., and Tsagris, M. (2004). Generation of transgenic potato plants highly resistant to potato virus Y (PVY) through RNA silencing. *Mol. Breeding* **14:**185–197.

Muller, G. W. and Rezende, J. A. M. (2004). Preimmunization: Application and Perspectives in Viral Disease Control. *In* "Diseases of Fruits and Vegetables, Diagnosis and Management" (S. A. M. H. Naqvi, ed.), pp. 361–396. Kluwer Academic Publishers, Dordrecht.

Nelson, R. S., Powell Abel, P., and Beachy, R. N. (1987). Lesions and virus accumulation in inoculated transgenic tobacco plants expressing the coat protein gene of tobacco mosaic virus. *Virology* **158:**126–132.

Nelson, R. S., McCormick, S. M., Delannay, X., Dube, P., Layton, J., Anderson, E. J., Kaniewskia, M., Proksch, R. K., Horsch, R. B., Rogers, S. G., Fraley, R. T., and Beachy, R. N. (1988). Virus tolerance, plant growth and field performance of transgenic tomato plants expressing coat protein from tobacco mosaic virus. *Biotechnology* **6:**403–409.

Nelson, R. S., Roth, D. A., and Johnson, J. D. (1993). Tobacco mosaic virus infection of transgenic *Nicotiana tabacum* plants is inhibited by antisense constructs directed at the 5′ region of viral RNA. *Gene* **127:**227–232.

Nejidat, A., and Beachy, R. N. (1990). Transgenic tobacco plants expressing a coat protein gene of *Tobacco mosaic virus* are resistant to some other tobamoviruses. *Mol. Plant-Microbe Interact.* **3:**247–251.

Niu, Q-W., Lin, S-S., Reyes, J. L., Chen, K-C., Wu, H-W., Yeh, S-D., and Chua, N-H. (2006). Expression of artificial microRNAs in transgenic *Arabidopsis thaliana* confers virus resistance. *Nature Biotechnol.* **11:**1420–1428.

Obbard, D. J., Gordon, K. H., Buck, A. H., and Jiggins, F. M. (2009). The evolution of RNAi as a defence against viruses and transposable elements. *Phil Trans. R. Soc. B* **364:**99–115.

Osbourn, J. K., Watts, J. W., Beachy, R. N., and Wilson, T. M. A. (1989). Evidence that nucleocapsid disassembly and a later step in virus replication are inhibited in transgenic tobacco protoplasts expressing TMV coat protein. *Virology* **172:**370–373.

Palukaitis, P and Zaitlin, M. (1984). A model to explain the "cross protection" phenomenon shown by plant viruses and viroids. *In* "Plant–Microbe Interactions" (T. Kosuge and E. W. Nester, eds.), pp. 420–429. McMillan, New York.

Powell Abel, P., Nelson, R. S., De, B., Hoffman, N., Rogers, S. G., Fraley, R. T., and Beachy, R. N. (1986). Delay of disease development in transgenic plants that express the tobacco mosaic virus coat protein gene. *Science* **232:**738–743.

Powell Abel, P., Stark, D. M., Sanders, P. R., and Beachy, R. N. (1989). Protection against *Tobacco mosaic virus* in transgenic plants that express Tobacco mosaic virus antisense RNA. *Proc. Natl. Acad. Sci. U. S. A.* **86:**6949–6952.

Praveen S., Ramesh, S.V., Mishra, A.K., Koundal, V., and Palukaitis, P. (2009). Silencing potential of viral derived RNAi constructs in *Tomato leaf curl virus*-AC4 gene suppression in tomato. *Trans. Res.* 10.1007/s11248-009-9291-y.

Prins, M., Laimer, M., Noris, E., Schubert, J., Wassenegger, M., and Tepfer, M. (2008). Strategies for antiviral resistance in transgenic plants. *Mol. Plant Pathol.* **88:**2638–2647.

Qu, F., Ye, X., and Morris, T. J. (2008). Arabidopsis DRB4, AGO1, AGO7, and RDR6 participate in a DCL4-initiated antiviral RNA silencing pathway negatively regulated by DCL1. *Proc. Natl. Acad. Sci. U. S. A.* **105:**14732–14737.

Ratcliff, F., Harrison, B. D., and Baulcombe, D. C. (1997). A similarity between viral defense and gene silencing in plants. *Science* **276:**1558–1560.

Ravelonandro, M. (2007). Transgenic virus resistance using homology-dependent RNA silencing and the impact of mixed virus infections. *In* "Biotechnology and Plant Disease Management" (Z. K. Punja, S. H. De Boer and H. Sanfaçon, eds.), pp. 374–394.

Register, J. C. III, and Beachy, R. N. (1988). Resistance to TMV in transgenic plants results from interference with an early event in infection. *Virology* **166:**524–532.

Register, J. C., and Beachy, R. N. (1989). Effect of protein aggregation state on coat protein-mediated protection against *Tobacco mosaic virus* using a transient protoplast assay. *Virology* **173:**656–663.

Sanders, P. R., Sammons, B., Kaniewski, W., Haley, L., Layton, J., Lavallee, B. J., Delannay, X., and Tumer, N. E. (1992). Field resistance of transgenic tomatoes expressing the *Tobacco mosaic virus* or *Tomato mosaic virus* coat protein genes. *Phytopathology* **82:**683–690.

Sanford, J. C., and Johnston, S. A. (1985). The concept of parasite-derived resistance —Deriving resistance genes from the parasite's own genome. *J. Theor. Biol.* **113:**395–405.

Schubert, J., Matoušek, J., and Mattern, D. (2004). Pathogen-derived resistance in potato to Potato virus Y—Aspects of stability and biosafety under field conditions. *Virus Res.* **100:**41–50.

Sharp, G. L., Martin, J. M., Lanning, S. P., Blake, N. K., Brey, C. W., Sivamani, E., Qu, R., and Talbert, L. E. (2002). Field evaluation of transgenic and classical sources of *Wheat streak mosaic virus* resistance. *Crop Sci.* **42:**105–110.

Shea, K (2009). University of Florida; Determination of nonregulated status for papaya genetically engineered for resistance to the papaya ringspot virus. Animal and Plant Health Inspection Service. *Federal Register* **74:**45163–45164 [http://edocket.access.gpo.gov/2009/pdf/E9-21092.pdf]

Sherwood, J. L. (1987). Mechanisms of cross-protection between plant virus strains. Plant resistance to viruses. *In* "Ciba Foundation Symposium 133" (D. Evered and S. Harnett, eds.), pp. 136–150. Wiley, Chichester.

Smith, N. A., Singh, S. P., Wang, M. B., Stoutjesdijk, P. A., Green, A. G., and Waterhouse, P. M. (2000). Total silencing by intron-spliced hairpin RNAs. *Nature* **407:**319–320.

Stanley, J., Frischmuth, T., and Ellwood, S (1990). Defective viral DNA ameliorates symptoms of geminivirus infection in transgenic plants. *Proc. Natl. Acad. Sci. U. S. A.* **87:**6291–6295.

Stone, R. (2008). China plans $3.5 billion GM crops initiative. *Science* **321:**1279.

Suzuki, J. Y., Tripathi, S., and Gonsalves, D. (2007). Virus-resistant transgenic papaya: Commercial development and regulatory and environmental issues. *In* "Biotechnology and Plant Disease Management" (Z. K. Punja, S. H. De Boer and H. Sanfaçon, eds.), pp. 436–461, CAB International, Cambridge, MA, USA.

Szittya, G., Silhavy, D., Molnar, A., Havelda, Z., Lowas, A., Lakatos, L., Banfalvi, Z., and Burgyan, J. (2003). Low temperature inhibits RNA silencing-mediated defence by the control of siRNA generation. *EMBO J.* **22:**633–640.

Tepfer, M. (2002). Risk assessment of virus-resistant transgenic plants. *Annu. Rev. Phytopathol.* **40:**467–491.

Tougou, M., Furutani, N., Yamagishi, N., Shizukawa, Y., Takahata, Y., and Hidaka, S. (2006). Development of resistant transgenic soybeans with inverted repeat-coat protein genes of soybean dwarf virus. *Plant Cell Rep.* **25:**1213–1218.

Tricoli, D. M., Carney, K. J., Russell, P. F., McMaster, J. R., Groff, D. W., Hadden, K. C., Himmel, P. T., Hubbard, J. P., Boeshore, M. L., and Quemada, H. D. (1995). Field evaluation of transgenic squash containing single or multiple virus coat protein gene constructs for resistance to *Cucumber mosaic virus*, *Watermelon mosaic virus 2*, and *Zucchini yellow mosaic virus*. *Biotechnology* **13:**1458–1465.

Vardi, E., Sela, I., Edelbaum, O., Livneh, O., Kuznetsova, L., and Stram, Y (1993). Plants transformed with a cistron of potato virus Y protease (NIa) are resistant to virus infection. *Proc. Natl. Acad. Sci. U. S. A.* **90:**7513–7517.

Voinnet, O. (2001). RNA silencing as a plant immune system against viruses. *Trends Genet.* **17:**449–459.

Voinnet, O. (2005). Induction and suppression of RNA silencing: Insights from viral infections. *Nature Rev. Genet.* **6:**206–221.

Voinnet, O. (2008). Post-transcriptional RNA silencing in plant–microbe interactions: A touch of robustness and versatility. *Curr. Opin. Plant Biol.* **11:**464–470.

Voinnet, O., Pinto, Y. M., and Baulcombe, D. (1999). Suppression of gene silencing: a general strategy used by diverse DNA and RNA viruses of plants. *Proc. Natl. Acad. Sci. U. S. A.* **96:**14147–14152.

Wang, Y., Bao, Z., Zhu, Y., and Hua, J. (2009). Analysis of temperature modulation of plant defense against biotrophic microbes. *Mol. Plant-Microbe Interact.* **22:**498–506.

Waterhouse, P. M., Graham, M. W., and Wang, M.-B. (1998). Virus resistance and gene silencing in plants is induced by double-stranded RNA. *Proc. Natl. Acad. Sci. U. S. A.* **95:**13959–13964.

Waterhouse, P. M., Smith, N. A., and Wang, M.-B. (1999). Virus resistance and gene silencing: Killing the messenger. *Trends Plant Sci.* **4:**452–457.

Waterhouse, P. M., Wang, M. B., and Lough, T. (2001). Gene silencing as an adaptative defense against viruses. *Nature* **411:**834–842.

Wesley, V., Helliwell, C. A., Smith, N. A., Wang, M. B., Rouse, D. T., Liu, Q., Gooding, P. S., Singh, S. P., Abbott, D., Stoutjesdijk, P. A., Robinson, S. P., Gleave, A. P., Green, A. G., and Waterhouse, P. M. (2001). Construct design for efficient, effective and high-throughput gene silencing in plants. *Plant J.* **27:**581–591.

Wintermantel, W. M., and Zaitlin, M. (2000). Transgene translatability increases effectiveness of replicase-mediated resistance to *Cucumber mosaic virus*. *J. Gen. Virol.* **81:**587–595.

Yang, H., Ozias-Akins, P., Culbreath, A. K., Gorbet, D. W., Weeks, J. R., Mandal, B., and Pappu, H. R. (2004). Field evaluation of *Tomato spotted wilt virus* resistance in transgenic peanut (*Arachis hypogaea*). *Plant Dis.* **88:**259–264.

Zaccomer, B., Cellier, F., Boyer, J. C., Haenni, A. L., and Tepfer, M. (1993). Transgenic plants that express genes including the 3′ untranslatable region of the turnip yellow mosaic virus (TYMV) genome are partially protected against TYMV infection. *Gene* **136:**87–94.

Zagrai, I., Capote, N., Ravelonandro, M., Ravelonandro, M., Cambra, M., Zagrai, L., and Scorza, R. (2008). Plum pox virus silencing of C5 transgenic plums is stable uner challenge inoculation with heterologous viruses. *J. Plant Pathol.* **90:** S1.63–S1.71.

CHAPTER 6

Genetically Engineered Virus-Resistant Plants in Developing Countries: Current Status and Future Prospects

D.V.R. Reddy*, **M.R. Sudarshana**†, **M. Fuchs**¶, **N.C. Rao**** and **G. Thottappilly**‡

Contents

Dedicated to the Late Nobel Laureate Dr. Norman Borlaug, father of the green revolution, for ardently supporting biotechnological approaches for increasing crop production and the improvement of crop quality.

* Venkata Villa, 8-2-283/B/1, Road #3 Banjara Hills, Hyderabad, 500034, India
† USDA-ARS, Department of Plant Pathology, University of California, One Shields Av., Davis, CA 95616, USA
¶ Department of Plant Pathology and Plant-Microbe Biology, Cornell University, New York State Agricultural Experiment Station, Geneva, NY 144566, USA
** Centre for Economic and Social Studies, Hyderabad 500016, India
‡ Sahrdaya College of Engineering and Technology, Kodaka, Trichur Dist., Kerala, 680684, India

Advances in Virus Research, Volume 75
ISSN: 0065-3527, DOI: 10.1016/S0065-3527(09)07506-X

Abstract

Plant viruses cause severe crop losses worldwide. Conventional control strategies, such as cultural methods and biocide applications against arthropod, nematode, and plasmodiophorid vectors, have limited success at mitigating the impact of plant viruses. Planting resistant cultivars is the most effective and economical way to control plant virus diseases. Natural sources of resistance have been exploited extensively to develop virus-resistant plants by conventional breeding. Non-conventional methods have also been used successfully to confer virus resistance by transferring primarily virus-derived genes, including viral coat protein, replicase, movement protein, defective interfering RNA, non-coding RNA sequences, and protease, into susceptible plants. Non-viral genes (R genes, microRNAs, ribosome-inactivating proteins, protease inhibitors, dsRNAse, RNA modifying enzymes, and scFvs) have also been used successfully to engineer resistance to viruses in plants. Very few genetically engineered (GE) virus resistant (VR) crops have been released for cultivation and none is available yet in developing countries. However, a number of economically important GEVR crops, transformed with viral genes are of great interest in developing countries. The major issues confronting the production and deregulation of GEVR crops in developing countries are primarily socio-economic and related to intellectual property rights, biosafety regulatory frameworks, expenditure to generate GE crops and opposition by non-governmental activists. Suggestions for satisfactory resolution of these factors, presumably leading to field tests and deregulation of GEVR crops in developing countries, are given.

I. INTRODUCTION

A. Need for genetically engineered crops in developing countries

In 2006, the United Nations estimated that the world population will reach 9.3 billion by 2050 (http://www.uniorg/esa/population/publications/wpp2006). World hunger is projected to reach a historic high in 2009 with 1.02 billion people (1/6th of population) going hungry every day, according to FAOs estimates (www.wfp.org/hunger). FAO also projects that global food production should increase more than 70% by 2050 to attain sufficiency, as production at the current rate will feed only 6.5 billion (FAO, 2008a). Increasing yields of staple food crops by adopting improved cultural methods and growing cultivars with traits that can enhance agronomic performance remains the most viable option to increase food production. Staple food crops such as maize, rice, and soybeans can be transformed for resistance to diseases, pests, and herbicides and for increased nutrition. In 2008, a record 13.3 million farmers across the planet planted genetically engineered (GE) crops (James, 2009). Globally (from 25 countries) the area under GE crops reached 125 million hectares in 2008, representing a 74-fold increase since 1996, when the first GE crop was released for commercial cultivation (James, 2009). The People's Republic of China, India, Mexico, and South Africa contributed nearly 40% of the total area under GE crops, the majority of which are insect-resistant or herbicide-tolerant crops (ISAAA, 2009). GE crops have the potential to overcome to a large extent the shortfall in food production (Zilberman *et al.*, 2004).

B. Economic losses caused by plant viruses

Annual crop losses due to plant diseases are estimated worldwide at $60 billion. Although losses caused by plant viruses alone are difficult to estimate, viruses are considered to be the second greatest contributor to yield loss, the foremost being fungi (Hsu, 2002). Losses estimated for major virus diseases are presented in Table 1. For example, $1.5 billion losses due to rice tungro disease are reported in South and Southeast Asia (Dai *et al.*, 2008). Viruses with a severe negative impact on agriculture for which actual economic losses are yet to be quantified are listed in Table 2. Indirect losses stemming from increased susceptibility of virus-diseased plants to other pathogens and abiotic stress factors can be substantial but are difficult to quantify.

Plant viruses are transmitted largely by arthropods that include aphids, whiteflies, and leaf or plant hoppers, in a persistent or non-persistent manner. The various options currently available to control

TABLE 1 Crop loss estimates for some economically important plant viruses in developing countries

Crop	Region/Country	Disease(s)	Losses	References
Banana	Worldwide	Streak	82%	Dahal *et al.* (2000)
Cassava	Africa	Cassava mosaic	1200–2300 m	Thresh *et al.* (1997)
Cassava	India and Sri Lanka	Mosaic	13–31%	Jeeva (1997)
Cardamom	India	Mosaic	70–100%	Roy *et al.* (2003)
Cereals (Barley, oat, and wheat)	Worldwide	Barley yellow dwarf and Cereal yellow dwarf viruses	11–12%	Henry and Adams (2003)
Cocoa	Africa	Swollen shoot	100%	Crowdy and Posnette (1947)
Cotton	Asia	Leaf curl	68–71%	Dasgupta *et al.* (2003)
Groundnut/peanut	Africa	Rosette	100 m	Reddy and Thirumala-Devi (2003)
Groundnut/peanut	India	Bud necrosis	89 m	Reddy *et al.* (1990)
Groundnut/peanut	India/W. Africa	Clump	38 m	Reddy *et al.* (2008)
Grapevine	France	Fanleaf	100 m	Fuchs (2008)
Maize	Africa	Streak	17–71%	Bosque-Perez (2000)
Pigeonpea	India	Sterility mosaic	300 m	Makkouk *et al.* (2003)
Potato	Worldwide	Leafroll	33–50%	Loebenstein and Manadilova (2003)
Potato	UK	Potato virus X, potato virus Y, and leaf roll	5.5 m	Fuchs (2008)

Potato	Worldwide	Potato X virus	10%	Loebenstein and Manadilova (2003)
Rice	South and SE Asia	Tungro	1500 m	Dai *et al*. (2008)
Rice	Africa	Yellow mottle	84–97%	Taylor (1989)
Rice	South America	Hoja blanca	25–50%	Calvert *et al*. (2003)
Soybean	Worldwide	Mosaic	8–25%	Hill (2003)
Sugarcane	Worldwide except Guyana and Mauritius	Mosaic	30–80%	Smith and Rott (2003)
Tomato	Tropical and subtropical regions	Yellow leaf curl	93–100% (losses in Jordan)	Antignus (2003)
Sweet potato	Worldwide	Sweet potato feathery mottle and sweet potato chlorotic stunt	50% in combined infections	Okada *et al*. (2001)

"m" in US$ in Millions or in percentage of loss from total yield

TABLE 2 Crop/virus combinations of economical importance in developing countries for which loss estimates are currently not available

Crop	Region/Country	Virus disease(s)	References
Banana	Africa, Australia, Asia, and the Pacific Islands	Bunchy top	Thomas *et al.* (2003)
Cassava	Kenya, Tanzania, Malawi, and Mozambique	Brown streak	Thottappilly *et al.* (2003)
Cassava	South America	Frogskin, severe stunting	Thottappilly *et al.* (2003)
Citrus	Worldwide	Tristeza	Lee and Bar-Joseph (2003)
Common bean	Worldwide	Mosaic due to potyviruses	Morales (2003)
Cowpea	Worldwide	Mosaic due to potyviruses	Hampton and Thottappilly (2003)
Cowpea	Africa, Asia, and South America	Golden mosaic	Hampton and Thottappilly (2003)
Cucurbits	Worldwide	Mosaic by several viruses	Lecoq (2003)
Faba bean, pea, chickpea, lentil	Worldwide	Leafroll	Makkouk *et al.* (2003)
Papaya	Worldwide	Ringspot	Fermin and Gonsalves (2003)
Yam	West Africa and West Indies	Mosaic	Atiri *et al.* (2003)

diseases caused by plant viruses include a set of cultural practices, application of insecticides/biocides to limit the multiplication and spread of vectors, and cultivation of resistant genotypes. Cultural practices include adjusting the date of planting relative to aerial vector flights, intercropping, spacing between individual plants and rows, mulching with reflective surfaces, and border cropping (Hull, 2009). In the case of fruit and ornamental crops, which are largely vegetatively propagated, planting certified virus-tested material can reduce or delay virus incidence. Virus-free stocks can be generated through tissue culture technologies, especially originating from meristem tips, in combination with heat treatment and clonal propagation (Razdan, 2003), followed by rigorous testing with sensitive virus detection methods to ensure freedom from viruses. However, planting virus-free material has limited success at controlling virus diseases under field conditions if sources of inoculum and vectors are present throughout the year. For example, the use of banana suckers free of *Banana bunchy top virus* (BBTV) by small-scale farmers in India did not delay or prevent the occurrence of bunchy top disease regardless if suckers were from certified material or from mother plants that did not show overt bunchy top symptoms (G. Thottappilly, unpublished).

Development of plant genotypes, which can resist or tolerate virus infection, even under low input farming practices, is by far the most effective way to control plant virus diseases (Fofana *et al.*, 2003). The most widely adopted conventional method consists of transferring and introgressing resistance genes from resistant germplasm by plant breeding techniques and subsequent rigorous selection of segregating lines under laboratory and field conditions. With the advent of marker-assisted selection, plant genotypes possessing the targeted genes are identified faster (Barone and Frusciante, 2007). Marker-assisted selection has recently been applied for rapid identification of promising maize lines resistant to *Maize streak virus* (MSV; Abalo *et al.*, 2009). Nonetheless, it is necessary to resort to non-conventional methods to produce GEVR crops due to some limitations of conventional breeding, including (i) long periods, often exceeding 5 to 10 years or more, needed for the transfer of virus resistance from natural sources of resistance to susceptible genotypes, this excludes the time needed to identify the sources of resistance, (ii) resistance largely location specific and often applicable to those virus isolates for which the resistance is evaluated, (iii) resistance often confined to narrow germplasm base, and (iv) occurrence of sterile hybrids in some crops (e.g., banana).

It should, however, be noted that durable resistance has been identified for many economically important viruses, such as *Peanut bud necrosis virus, Groundnut rosette viruses, Cassava mosaic geminiviruses, Cassava brown streak virus, Cacao swollen shoot virus, Pigeonpea sterility mosaic virus, Rice tungro bacilliform virus, Rice tungro spherical virus, Rice yellow mottle virus, Maize streak virus, Sugarcane mosaic virus, Citrus tristeza virus, Cotton leaf curl virus, and Tomato yellow leaf curl virus*, that are prevalent in developing countries (Loebenstein and Carr, 2006; Loebenstein and Thottappilly, 2003).

In this article, we review options for developing GEVR plants, list GEVR crops of interest to developing countries, and discuss factors that limit their development and release in developing countries.

II. OPTIONS FOR DEVELOPING GENETICALLY ENGINEERED RESISTANCE TO VIRUSES AND THEIR VECTORS

The introduction and expression of nucleotide sequences of viral and non-viral origin in plants to confer virus resistance were made possible by gene splicing technologies, development of promoters that ensure constitutive expression of genes, and advances in techniques that facilitate transformation of a range of crop plants. It has been nearly 25 years since the first report on the development of tobacco plants resistant to *Tobacco mosaic virus* (TMV) (Powell-Abel *et al.*, 1986). Since then numerous reports have documented the successful production of plants with durable resistance to a number of different viruses (Baulcombe 1996; Beachy *et al.*, 1990; Dasgupta *et al.*, 2003; Goldbach *et al.*, 2003; Prins *et al.*, 2008; Sudarshana *et al.*, 2006; Wilson, 1993). We give a brief account of the anti-viral genes successfully deployed to genetically engineer plants for virus resistance. A short account of genes that can be used to silence plant host genes and ways to interfere with virus transmission by insect vectors are also included.

A. Anti-viral genes of viral origin

1. Protein-mediated resistance

Coat protein genes. The most widely used genes to engineer virus resistance in plants were derived from coat protein (CP) genes, either as full length or truncated constructs. The degree of protection conferred by CP genes varied from near immunity to delay in the expression of overt symptoms. In some cases, the resistance was broad (Beachy *et al.*, 1990; Lomonossoff, 1995). All of the currently commercially released virus-resistant crops are developed through incorporation of CP genes. The molecular mechanisms underlying the CP-mediated resistance are not fully understood and appear to differ among different viruses (Prins *et al.*, 2008). Resistance is attributed to viral protein or its RNA or co-existence of both.

Replication associated proteins. Many single-stranded DNA viruses are economically important in developing countries. Replication of these viruses requires interaction between replication associated protein (Rep) and host polymerases. Rep is indispensable for virus replication (Elmer *et al.*, 1988). In Section III, examples of the successful production of GEVR plants to the geminiviruses *African cassava mosaic virus* (ACMV) and TYLCV using the Rep approach are provided.

RNA-dependent RNA polymerase genes. Tobacco plants expressing part of the viral RNA-dependent RNA polymerase (RdRp or replicase) showed resistance to TMV (Golemboski *et al.*, 1990) and the resistance was strain specific. Expression of the TMV 54-kDa protein corresponding to the carboxyl end of the 183-kDa RdRp was essential to confer resistance (Carr *et al.*, 1992). Similarly expression of full-length 54 kDa protein or 30% of it was essential to induce resistance to another tobamovirus, *Pepper mild mottle virus* (Tenllado *et al.*, 1995). Later studies attributed this resistance to RNA silencing (Marano and Baulcombe, 1998). Replicase genes derived from potex, poty, alfamo, and cucomoviruses have been used successfully to induce resistance (Anderson *et al.*, 1992; Audy *et al.*, 1994; Braun and Hemenway, 1992; Carr *et al.*, 1994; Wintermantel and Zaitlin, 2000). It is apparent that a high degree of resistance can be induced to several plant viruses through transfer of native or truncated forms of replicase genes.

Movement protein genes. Tobacco plants GE with a mutant movement protein (MP) of TMV blocked not only the movement of TMV, but also of several other plant viruses including some with a DNA genome (Lapidot *et al.*, 1993). Interestingly, expression of native MP supported infection by movement deficient virus strain (Deom *et al.*, 1987). Movement of potex-, carla-, hordei-, and furoviruses is regulated by a 13-kDa protein, which is part of a triple gene block protein complex. Expression of mutant forms of the 13 kDa protein conferred resistance to the homologous virus, as well as some that are dependent on the triple gene block protein complex for their movement (Seppänen *et al.*, 1997).

Protease genes. The RNA genome of virus species that belong to the picornavirus super family is expressed by proteolytic cleavage of a polyprotein, which is processed by a virus encoded protease (Ryan and Flint, 1997). For potyviruses, the protease is referred to as NIa (Urcuqui-Inchima *et al.*, 2001). Plants expressing the viral protease domain of NIa exhibited a high degree of resistance to virus infection, presumably through interference of polyprotein processing into functional products essential for virus replication (Maiti *et al.*, 1993). Integrity of the protease active site is essential to induce resistance but protease activity alone was not adequate to elicit protease-mediated resistance in potato to PVY (Mestre *et al.*, 2003).

2. RNA-mediated resistance

Resistance to viruses in plants can be mediated through RNA silencing (Prins *et al.*, 2008). Expression of translatable or untranslatable RNA and anti-sense RNA corresponding to CP cistron of two potyviruses in tobacco plants conferred a high degree of resistance to mechanical inoculation as well as infection via aphids (Lindbo and Dougherty, 1992; van der Vlugt *et al.*, 1992). This type of resistance can also be incorporated to protect plants against RNA and DNA virus infection through the expression of short hairpin RNAs (shRNAs) (Lucioli *et al.*, 2003; Prins *et al.*, 2008). The pathway leading to the generation of virus-specific small-interfering RNAs (siRNAs) (Hamilton and Baulcombe, 1999) and systemic silencing have been discussed in Chapter 2.

B. Anti-viral genes of non-viral origin

1. Plant-derived *R* genes and micro RNAs

Traditionally, plant pathologists, germplasm scientists and breeders have worked together in identifying sources of plant virus resistance (*R*) genes in cultivated and wild species. Wherever possible, *R* genes have been bred and introgressed into cultivated varieties. The process can take 5 to 10 years or more before a genotype with desirable traits is made available for commercial release. Several of the plant virus resistance genes have been cloned and, as transgenes, have shown resistance to plant viruses. The *N* gene from the tobacco variety Samsun NN was the first gene to be cloned and this gene conferred resistance to TMV in tobacco and tomato (Whitham *et al.*, 1994, 1996). The majority of virus-specific *R* genes cloned so far are single dominant resistance genes and their molecular architecture is similar to plant resistance genes against bacteria, fungi, and nematodes (Goldbach *et al.*, 2003; McHale *et al.*, 2006). In addition to *R* genes, plants also have recessive resistance genes, which confer impaired susceptibility to virus infection (Maule *et al.*, 2007). For additional information on dominant and recessive resistance genes and their products See Chapters 1 and 4.

Plants also encode microRNAs (miRNAs) of 24 nt in length which regulate the abundance of cognate mRNAs by guiding their cleavage (Niu *et al.*, 2006). These miRNAs are generated by processing of pre-miRNA prescursors by dicer DCL1 in *Arabidopsis thaliana* (Xie *et al.*, 2004). Niu *et al.* (2006) modified a plant miRNA precursor, miR159 from *A. thaliana*, to express artificial miRNAs, leading to suppression of viral genes, P69 from *Turnip yellow mosaic virus* (TYMV) and HC-Pro from *Turnip mosaic virus* (TuMV). Transgenic *A. thaliana* plants expressing modified miRNAs showed resistance to TYMV or TuMV and plants expressing both miRNAs were resistant to the two viruses. Artificial miRNA resistance was found to be more desirable than that obtained with shRNAs (Qu *et al.*, 2007). However, it is premature to conclude that this approach is superior to incorporating viral transgenes.

2. Ribosome-inactivating proteins

Ribosome-inactivating proteins (RIPs), that contain two polypeptide chains A and B, are catalytic enzymes that deglycosylate a specific base in the 28S rRNA of eukaryotes. Chain A is the catalytic domain and chain B binds to galactose. The majority of plant RIPs are type 1 that contain only chain A. Pokeweed anti-viral protein (PAP) is a RIP known to inhibit virus multiplication. Transgenic *N. tabacum* plants that expressed PAP showed broad-spectrum resistance to several viruses, including *Cucumber mosaic virus* (CMV), *Potato virus Y* (PVY), and *Potato virus X* (PVX), when challenged by mechanical inoculation as well as via aphid transmission (Lodge *et al.*, 1993). This approach has been shown also to work against a plant DNA virus (Hong *et al.*, 1996). PAP can degrade 23S rRNA (Chaddock *et al.*, 1994) and 28S rRNA (Tumer *et al.*, 1997) of prokaryotes. The mechanism of PAP-mediated resistance to virus infection was initially thought to be due to inactivation of the 28S rRNA. However, a deletion mutant of PAP in the C-terminal domain of 28S rRNA that lacked depurination ability still exhibited virus resistance in transgenic plants (Tumer *et al.*, 1997). The mechanism underlying the resistance in GE plants expressing PAP or RIP type 1 is currently not well understood.

3. Protease inhibitors

Members of several economically important plant viruses that belong to the families *Comoviridae*, *Potyviridae*, and *Closteroviridae* encode for *cis* and/or *trans*-acting proteases to release individual proteins from a polyprotein encoded by their genomic RNAs. Protease inhibitors can be expected to interfere with viral proteases. Tobacco plants expressing a rice cystatin (a cysteine proteinase inhibitor) were resistant to *Tobacco etch virus* (TEV) and PVY but not to TMV (Gutierrez-Campos *et al.*, 1999). Presumably, the product of the transgene could confer resistance against those viruses that are dependent on proteases for genome expression and replication. We are not aware of any report on the successful production of GEVR food crops using this approach.

C. Anti-viral genes from eukaryotes

1. Expression of dsRNAse and RNA modifying enzymes

Plant RNA viruses and viroids replicate through a dsRNA intermediate. One of the yeast genes *pac* 1 encodes for a ribonuclease capable of degrading dsRNA. Watanabe *et al.* (1995) isolated this gene from *Shizosaccharomyces pombe* and expressed it in *N. tabacum* cv. Xanthi-nc plants. These plants showed a decrease in lesion numbers when challenged by *Tomato mosaic virus* and a delay in the development of symptoms when inoculated with CMV and PVY. Further experiments conducted on potato plants expressing *pac* 1 demonstrated that *Potato spindle tuber viroid* infection was suppressed in these plants, and the tubers were also free of the viroid (Sano *et al.*, 1997).

In mammals, the interferon system provides a broad-spectrum anti-viral response. During the course of viral infection, interferons induce synthesis of additional proteins that affect virus multiplication. Among these, 2′–5′ oligoadenylate synthetase (2–5Ase) is a key enzyme which upon activation by the presence of dsRNA polymerizes ATP to produce 2′–5′ oligoadenylates, which in turn activates RNAse L, an endoribonuclease, that degrades both viral and cellular RNAs. Potato plants expressing a rat cDNA encoding the 2–5Ase were protected from PVX infection under field conditions (Truve *et al.*, 1993). Some of the GE potato lines were found to be superior in resisting virus infection to those expressing a PVX *CP* transgene. Expression of a human 2–5Ase in tobacco also protected plants against CMV and TMV (Ehara *et al.*, 1994). When tobacco plants with human RNase L were crossed with those GE with 2–5Ase, offsprings carrying both transgenes were more resistant to CMV and PVY infection compared to parent GE lines carrying either RNAse L or 2–5Ase alone (Ogawa *et al.*, 1996).

2. Plantibodies

Ever since the development of antibodies raised against purified plant virus preparations, expression of antibodies in plants was presumed to provide protection against virus infection. *N. benthamiana* containing a single chain variable fragment (scFv) specific to the CP of *Artichoke mottled crinkle virus* showed reduction in disease incidence and delay in symptom expression (Tavladoraki *et al.*, 1993). Several other reports have been published on the successful expression of scFvs specific to CPs of several plant viruses (Prins *et al.*, 2008). A novel approach based on expressing scFvs to a virus replicase produced plants with broad resistance to *Cucumber necrosis virus* and *Tomato bushy stunt virus* (Family *Tombusviridae*; Genus: *Tombusvirus*), *Red clover necrotic mosaic virus* (Family *Tombusviridae*; Genus: *Dianthovirus*), and *Turnip crinkle virus* (Family *Tombusviridae*; Genus: *Carmovirus*) (Boonrod *et al.*, 2004). A logical extension of this approach would be to test scFvs specific to the GDD box of RdRp to achieve broad-spectrum protection against plant RNA viruses. Fomitcheva *et al.* (2005) expressed MAbs to GDD domains of RdRp of *Barley yellow dwarf virus* (BYDV) and NIb of PVY. A scFv clone capable of detecting both antigens have been identified (Fomitcheva *et al.*, 2005) but GEVR plants expressing this scFv clone are yet to be produced.

3. Silencing of host genes involved in virus multiplication

Several host genes are actively involved in virus multiplication. Mutations of *A. thaliana* genes *TOM1* and *TOM3* abolished completely the multiplication of two strains of TMV (Yamanaka *et al.*, 2002). When homologs of these genes in *N. tabacum* cv. Samsun were silenced by RNA interference, plants were resistant to several tobamoviruses (Asano *et al.*, 2005). The loss of two host gene products apparently had no effect on the growth of silenced plants. Interestingly, this resistance was limited to tobamoviruses, but not to a CMV strain, indicating differences in the selection of host proteins utilized by viral RdRps. Silencing of *TOM1* and *TOM3* homologs in *N. benthamiana* also prevented tobamovirus multiplication (Chen *et al.*, 2007). Conceivably this approach can be applied to any plant virus, given the host genes to be silenced are identified and characterized. Diaz-Pendón *et al.* (2004) and Maule *et al.* (2007) have listed several host factors that influence infection by different viruses, providing opportunities to apply this strategy.

D. GE resistance to arthropod vectors or transmission of plant viruses

Another approach to control plant virus spread is through the development of resistance to insect vectors. GE rice lines expressing a lecithin gene caused significant reductions in the survival, fecundity, and development of the brown plant hopper, *Nilaparvatha lugens* (Rao *et al.*, 1998). In addition to being a vector of viruses that cause grassy stunt, ragged stunt, and tungro in rice, *N. lugens* is a major pest of rice.

Despite a vast amount of research on RNAi-mediated silencing targeted to viruses this technology is yet to be fully exploited to control plant virus vectors because of the inability of siRNA made by plants to trigger silencing in insects. A breakthrough in this regard came recently when silencing effects of dsRNA in artificial diet containing gossypol were studied in *A. thaliana*. siRNA made by plants was unable to trigger silencing in insects while shRNA was able to turn on silencing in insects artificially fed on diet spiked with shRNA targeted to insect genes (Mao *et al.*, 2007). shRNA-mediated silencing in agricultural crops is yet to be exploited for engineering virus resistance through controlling vectors.

Another protein of interest is a homolog of the GroEL protein, called symbionin. This protein is produced by bacterial endosymbionts of insects, including aphid and whitefly vectors of plant viruses. Symbionin has the capability to bind virions of persistently transmitted, circulative viruses. These viruses are dependent on this protein for, their safe passage through the insect hemolymph and their subsequent secretion by the salivary glands (Czosnek *et al.*, 2002). A *GroEL* gene obtained from whiteflies (*Bemisia tabaci*) conferred tolerance to TYLCV and CMV when expressed in transgenic *N. benthamiana* (Edelbaum *et al.*, 2009). Tolerance of plants expressing GroEL gene to CMV was unexpected because the virus is non-persistently transmitted. A GroEL homologue from *B. tabaci* was able to bind geminate and icosahedral RNA viruses (Akad *et al.*, 2004). The potential of GEVR plants expressing GroEL is yet to be fully exploited.

III. EXAMPLES OF SUCCESSFUL PRODUCTION OF GENETICALLY ENGINEERED CROPS WITH VIRUS RESISTANCE IN DEVELOPING COUNTRIES

A. GEVR crop plants with potential for release in developing countries

Many crops have been engineered for virus resistance. However, to date, no GEVR cultivars have been released by governmental agencies or the private sector for general cultivation in developing countries. GEVR crops tested under laboratory and field conditions with potential for future release in developing countries are listed in Table 3. Some of these crops and the targeted viruses are discussed below.

1. Maize streak disease

Maize is by far the most important staple food crop in Africa contributing 50% to the calorific intake of the population (http://faostat.fao.org).

TABLE 3 Engineered virus resistant crop plants with potential for release in developing countries

Crop plant	Virus	References
Cassava	*African cassava mosaic virus* (*Begomovirus*)	Fofana *et al.* (2003); Patil and Fauquet (2009)
Cereals	*Barley yellow dwarf virus* (*Luteovirus*)	Wang *et al.* (2000); Henry and McNab (2002)
Citrus	*Citrus tristeza virus* (*Closterovirus*)	Febres *et al.* (2008)
Maize	*Maize steak virus* (*Mastrevirus*)	Shepherd *et al.* (2007)
Cucumber	*Cucumber mosaic virus* (*Cucumovirus*)	Fuchs (2008)
Papaya	*Papaya ring spot virus* (*Potyvirus*)	Fuchs and Gonsalves (2007)
Potato	*Potato virus X* (*Potexvirus*), *Potato virus Y* (*Potyvirus*), *Potato leafroll virus* (*Polerovirus*)	Dasgupta *et al.* (2003)
Rice	*Rice stripe virus* (*Tenuivirus*)	Hayakawa *et al.* (1992)
Rice	*Rice yellow mosaic virus* (*Sobemovirus*)	Pinto *et al.* (1999)
Rice	*Rice Tungro* viruses (*Tungrovirus*)	Dai *et al.* (2008)
Sweet potato	*Sweet potato feathery mottle virus* (*Potyvirus*)	Kreuze *et al.* (2009)

Maize streak disease caused by the geminivirus MSV is a major constraint for maize production throughout sub-Saharan Africa (Bosque-Perez, 2000; Efron *et al.*, 1989; Gordon and Thottappilly, 2003; Rybicki and Peterson, 1999; Thottappilly *et al.*, 1993). Yield losses of up to 100% are known to occur in case of early infections. Maize expressing a mutated MSV Rep gene showed significant delay in symptom development and decreased symptom severity. Additionally GE plants exhibited higher survival rates compared to non-GE plants (Shepherd *et al.*, 2007). Resistance was shown to be stably inherited until the T3 generation and to be transferable by conventional breeding to elite maize breeding lines. However, durability of resistance to MSV remains to be assessed following vector-mediated inoculations under field conditions at various locations in Africa.

2. Cassava mosaic disease

Cassava mosaic disease is a major constraint for cassava cultivation throughout Africa and in the Indian subcontinent. In Africa and its adjacent islands (Fauquet and Fargette, 1990; Thottappilly *et al.*, 2003; Thresh and Fargette, 2001), seven geminiviruses, *African cassava mosaic virus* (ACMV), *East African cassava mosaic virus* (EACMV), *South African cassava mosaic virus*, *East African cassava mosaic Cameroon virus*, *East African cassava mosaic Malawi virus*, *East African cassava mosaic Kenya virus*, and *East African cassava mosaic Zanzibar virus*, have been shown to cause cassava mosaic disease (Fauquet *et al.*, 2008; Patil and Fauquet, 2009). In Uganda and neighboring countries, a new and severe virus variant was detected and identified. This variant resembled EACMV from which most of its genome was derived, except for the core region of the CP which was identical to that of ACMV (Harrison *et al.*, 1997). In India, two begomoviruses, *Indian cassava mosaic virus* (Hong *et al.*, 1993) and *Sri Lankan cassava mosaic virus* (Saunders *et al.*, 2002), cause cassava mosaic disease.

Several laboratories are currently working on engineered resistance to geminiviruses in cassava. Resistance was achieved using defective interfering DNA or a full-length coding sequence of the *Rep* gene and also by using a mutant of the putative NTP binding domain of the Rep gene (Fofana *et al.*, 2003). Zhang *et al.* (2005) demonstrated that resistance to ACMV can be achieved with high efficiency by expressing anti-sense RNAs against viral mRNAs encoding Rep, replication enhancer protein (Ren), and transcription activator protein (Trap). Vanderschuren *et al.* (2007) suggested a natural RNA silencing mechanism that targets ACMV through production of virus-derived siRNAs of 21–24 nt in length expressing dsRNA cognate to the viral promoter and common region. Expression of resistance was in the form of attenuation of cassava mosaic disease symptoms and reduced accumulation of ACMV DNA in GE plants. Despite the impressive progress made on the characterization of viruses that cause cassava mosaic disease and the availability of technologies to generate GEVR cassavas, GEVR cassavas are yet to be released for general cultivation in Africa. Reports on the consistent performance of ACMV-resistant GE cassava plants, tested under natural infection conditions in Africa, are lacking. Cassava varieties which are not readily infected by ACMV are available. Under high disease pressure they show mild symptoms (Thottappilly *et al.*, 2003). These varieties are ideal for incorporating GE resistance to ACMV.

Initiation of a new five-year (2009–2014) USDA project on the development and deployment of transgenic cassava-resistant ACMV and *Cassava brown streak virus* for cultivation by marginal farmers in Africa will help, hopefully mitigating the losses due to these viruses (www.ars.usda.gov/research/projects/projects.htm?ACCN_NO=416889).

3. Sweet potato feathery mottle virus disease

Several viruses have been shown to cause economic losses to sweet potato (*Ipomaea batatas*), among which, the aphid transmitted *Sweet potato feathery mottle virus* (SPFMV; *Family Potyviridae*) and whitefly transmitted *Sweet potato chlorotic stunt virus* (SPCSV; *Family Closteroviridae*) occur worldwide (Loebenstein *et al.*, 2009). In Africa, SPFMV in combination with SPCSV causes severe crop losses while SPFMV alone may cause only minor damage. CP-mediated resistance to SPFMV was introduced into several African varieties in addition to the widely grown cultivar CPT-560 by the Kenya Agricultural Research Institute with technical cooperation from the Monsanto company (Odame *et al.*, 2001). When these lines were evaluated under field conditions in Kenya, they were found to be susceptible to SPFMV when co-infected with SPCSV (New Scientist, 2004). Although GE sweet potato plants expressing rice cysteine proteinase inhibitor showed resistance to SPFMV but susceptibility to SPFMV was restored when co-infected with SPCSV (Cipriani *et al.*, 2001; Kreuze *et al.*, 2009). Therefore, it was essential to design a transgene that expressed a SPCSV-homologous transcript that formed a double-stranded structure and conferred high resistance to SPFMV (Kreuze *et al.*, 2008, 2009). Many transgenic lines showed low SPCSV concentration without overt symptoms. However, low concentration of SPCSV in transgenic plants was sufficient to breakdown the high levels of resistance to SPFMV (Kreuze *et al.*, 2009). Cuellar *et al.* (2009) have recently reported that transformation of SPFMV-resistant sweet potato with double-stranded RNA specific class I RNA endoribonuclease III (RNase 3) of SPCSV resulted in resistance breakdown to SPFMV. RNase 3 has been shown to contribute to suppression of anti-viral defense and mediated viral synergism in GE plants with several unrelated viruses to SPFMV. It is therefore essential to develop GE sweet potato plants in which replication of SPCSV is inhibited to the extent that it will not permit the replication of other viruses that infect sweet potato plants.

4. Barley yellow dwarf disease

Barley yellow dwarf is an economically important disease of cereal crops caused by the luteovirus BYDV (Henry and McNab, 2002). Six serotypes of BYDV have been described. Luteoviruses described under BYDV comprise two subgroups: BYDV-PAV, BYDV-MAV, and BYDV-ORV belong to subgroup 1 and *Cereal yellow dwarf virus* (CYDV) belongs to subgroup 2. More than 100 monocotyledonous species are susceptible to the viruses that belong to the two subgroups. Through conventional breeding, supported by marker-assisted selection, it is possible to develop BYDV-resistant wheat and barley cultivars. However, the resistance was found to be location specific (Kosova *et al.*, 2008). A single copy of a virus-derived transgene encoding a hairpin RNA sequence provided immunity to BYDV-PAV in barley (Wang *et al.*, 2000). Since the advent of cereal transformation, engineered resistance to BYDV has been explored (Dupre *et al.*, 2002).

5. Tomato yellow leaf curl disease

Tomato yellow leaf curl disease is one of the most serious constraints on tomato production (Moriones and Navas-Castillo, 2000). It is widely distributed in Asia, sub-Saharan Africa, the Caribbean Islands, Australia, and the USA. Tomato yellow leaf curl disease is caused by at least 57 different geminivirus species of which *Tomato yellow leaf curl virus* (TYLCV) belongs to the genus *Begomovirus*, family *Geminiviridae*, is a major contributor. Among TYLCV isolates two species with a single genomic component have been extensively studied: TYLCV and *Tomato yellow leaf curl Sardinia virus* (TYLCSV) (Abhary *et al.*, 2007).

Resistance incorporated through conventional breeding methods was shown to break down if plants were infected at early stages of growth and under high-infection pressure (Lapidot and Friedmann, 2002). Therefore incorporation of resistance by transforming tomato plants, utilizing the *Rep* gene, has been the most widely adopted approach (Lucioli *et al.*, 2003). A protein-mediated resistance was described in tomato plants with a gene encoding a truncated TYLCSV Rep of 210 amino acids. However, resistance in some cases was found to be unstable due to transgene silencing (Lucioli *et al.*, 2003). A similar construct, having a further deletion at the C-terminus and encoding the first 129 amino acids of the protein, induced resistance to TYLCV (Antignus *et al.*, 2004) and TYLCSV (Prins *et al.*, 2008). Tomato plants transformed with *Rep* gene sequences in anti-sense orientation showed high level of resistance to an Indian isolate of TYLCV (Shelly *et al.*, 2005). Broad-based resistance to TYLCV was achieved by using a chimeric construct composed of conserved sequences derived from the replicase active site, transcription enhancer, promoter, and silencing suppressor. However, the effectiveness of these GE plants is yet to be assessed under high disease pressure in the field (Polston and Hiebert, 2007). Recently, expression of a TYLCV scFv protected *N. benthamiana* plants from infection. This research is significant because it was the first instance of successful application of recombinant-antibody mediated resistance to a plant DNA virus (Safarnejad *et al.*, 2009). The effectiveness of this approach in tomato plants is yet to be reported.

It is apparent that some very effective transgene constructs coding Rep protein have been found to be suitable for incorporating resistance against TYLCV and TYLCSV into plants (Yang *et al.*, 2004). To our knowledge none of the transformed plants containing the genetic constructs described here have been released for general cultivation despite the pressing need to reduce crop losses due to tomato yellow leaf curl disease.

6. Banana bunchy top disease

Banana is an economically important crop grown in more than 120 countries. Among all the fruit crops, on a global scale, banana occupies the fourth place in terms of the total cultivated area. Its global production exceeds 104 million tons, 87% of which is produced by marginal farmers (http://r4dreview.org/2008/09/banana-facts). Banana bunchy top is the most devastating disease of banana worldwide (Presley and George, 1999). It is widespread in Southeast Asia, the Philippines, Taiwan, Hawaii, and the South Pacific islands, and in parts of India and Africa (Thomas *et al.*, 2003). The disease is caused by BBTV from the family *Nanoviridae*, genus *Babuvirus* (Büchen-Osmond, 2007). The aphid *Pentalonia nigronervosa* transmits BBTV. None of the commercially grown banana varieties is resistant to BBTV and generating disease-resistant bananas by conventional breeding is difficult because sterile progeny are produced from commercially grown hybrids (Sahijram, 2004). Therefore, the development of transgenic resistance is the most viable option to confer resistance to bunchy top disease. BBTV is a single-stranded DNA virus and a Rep protein is essential for its replication by host enzymes. Constructs from BBTV Rep gene, which can suppress significantly replication of viral DNA in banana embryonic cell suspensions, have been prepared (Tsao, 2008). We expect that these constructs will be used in the near future to transform bananas.

7. Rice yellow mottle disease

The disease caused by *Rice yellow mottle virus* (RYMV) is a serious problem in Africa. Rice transformed with a transgene encoding the replicase of RYMV showed complete suppression of virus multiplication as a result of post-transcriptional gene silencing (Pinto *et al.*, 1999).

Although GE resistance was overcome by several RYMV isolates, these virulent isolates seldom occur under field conditions (Sorho *et al.*, 2005). Field trials are needed to evaluate the plant reaction to RYMV under natural infection conditions.

B. Deregulated GEVR crops

In addition to the above-mentioned GEVR crops of interest to developing countries, other crops that are commercially available in the USA and the People's Republic of China could be potentially deployed in developing countries to help mitigate the impact of devastating virus diseases.

1. Papaya plants resistant to *Papaya ringspot virus*

Papaya (*Carica papaya*) is widely grown in tropical and subtropical lowland regions. Among the developing countries, India, Indonesia, Mexico, Nigeria, and Thailand are the largest papaya growing countries. *Papaya ringspot virus* (PRSV) is currently recognized as the most economically important virus that infects papayas (Fermin and Gonsalves, 2003). PRSV-resistant papaya is the first example of successful commercial application of GE technology in a fruit crop (Gonsalves, 1998). GEVR papayas were produced through integration of the entire or a truncated *CP* gene of PRSV. In extensive field tests papaya were resistant to PRSV and produced fruit yields comparable to uninfected conventional plants.

In the USA, GEVR papayas were deregulated in 1998. Cultivars derived from deregulated GEVR papayas are widely grown in Hawaii. Impact of PRSV-resistant cultivars, Rainbow and SunUp, on papaya production and economic returns are well documented (Fuchs and Gonsalves, 2007). In 2006, PRSV-resistant cultivars were planted on more than 90% of the total production land in Hawaii (Johnson *et al.*, 2007). GEVR papayas were also deregulated in The People's Republic of China (Karplus and Deng, 2008). Efforts are ongoing in several developing countries, including Brazil, Jamaica, Mexico, Thailand, the Philippines, and Venezuela to engineer resistance to PRSV in local papaya cultivars. In Thailand, the technology was tested successfully in multi-location field trials. However, deregulation of GEVR papayas was prevented in large part due to efforts of non-governmental organizations such as Greenpeace, which destroyed papaya plants in experimental field sites. This is a good example of how activists from wealthy countries have ruined the most sincere efforts by the Thai government and governments of other developing nations to introduce badly needed GEVR crops (Davidson, 2008). Despite the economic importance of PRSV in India, Nigeria, and Indonesia, GEVR papayas are yet to be released for general cultivation by commercial and marginal farmers. For more on this topic the reader is referred to Chapter 5.

2. Summer squash cultivars resistant to CMV, Zucchini yellow mosaic virus (ZYMV), and Watermelon mosaic virus (WMV)

Transgenic summer squash *Cucurbita pepo* cultivars resistant to ZYMV, WMV, and/or CMV were developed by transformation with the appropriate viral *CP* gene sequences (Tricoli *et al.*, 1995). Squash resistant to ZYMV and WMV were deregulated in 1996 and squash resistant to CMV, ZYMV, and WMV were deregulated in 1998 in the USA. Deregulated GEVR summer squash have also been used as parents in conventional breeding to develop 11 virus-resistant summer squash cultivars. The adoption rate of GEVR summer squash was estimated to 22% across the country in 2006 with a net benefit of $24 million (Johnson *et al.*, 2007). There are no records of introduction of GEVR summer squash in developing countries, in spite of several attempts in Africa.

3. Sweet pepper and tomato cultivars resistant to CMV

In the People's Republic of China, tomato and sweet pepper (*Capsicum* sp.) resistant to CMV were released (James, 2009; Stone, 2008). However, the performance of the transgenic tomato and sweet pepper under field conditions did not differ significantly from the comparable non-GE cultivars. As a result, investment in commercial production of these transgenic lines was discontinued in the People's Republic of China (Karplus and Deng, 2008).

IV. FACTORS LIMITING THE INTRODUCTION AND CULTIVATION OF GENETICALLY ENGINEERED VIRUS-RESISTANT CROP PLANTS IN DEVELOPING COUNTRIES WITH EMPHASIS ON SOCIO-ECONOMIC ISSUES

Prospects for incorporating durable transgenic virus resistance into desirable plant genotypes have increased enormously with the discovery of a range of viral and non-viral genes (Section II). PRSV-resistant papayas were deregulated in 1998 and currently more than 90% of the traditional papaya growing area in Hawaii is under cultivation with GE PRSV-resistant papayas (Johnson *et al.*, 2007).

Papayas from GEVR plants have been consumed in Hawaii and continental USA for the last decade. There are no reports of adverse effects arising on human health from consumption of these papayas. Numerous tests on papayas from GEVR plants did not reveal any differences in allergenicity between GE and non-transgenic fruit. Additionally, tests done on GEVR summer squash resistant to CMV,

WMV, and ZYMV did not show properties that can be attributed to allergens (Fuchs and Gonsalves, 2007). Similarly, consumption of papaya fruits infected with a mild PRSV strain through cross-protection did not cause any ill effects in humans (Yeh and Gonsalves, 1994). However, no GEVR plant has been deregulated in developing countries with the exception of the People's Republic of China (Section III).

Potential environmental safety issues related to the cultivation of GEVR plants and important social and ethical issues, which are likely to limit their introduction in developing countries, are discussed in this section. The environmental risks arising from recombination, transmission of transgenes through pollen, and effect on non-target organisms were assessed extensively. These were reviewed and discussed recently (Dale *et al.*, 2002; Fuchs and Gonsalves, 2007; Tepfer, 2002). For example, the emergence of recombinant viruses was not detected in transgenic plum trees expressing the CP gene of *Plum pox virus* (PPV) that were grown over a period of 10 years in Spain and Romania (Capote *et al.*, 2007; Fuchs *et al.*, 2007; Zagrai *et al.*, 2008). Additionally no detectable difference was noticed in the number and type of aphids that visited GE and non-GE plum trees in an experimental orchard in Spain (Capote *et al.*, 2007). Also, challenging PPV-resistant plums with three unrelated viruses did not lead to a breakdown of the engineered resistance to PPV (Zagrai *et al.*, 2008). Taken together, the studies show that risks to the environment and human health, which were of serious potential concern two decades ago, are no longer considered to be constraints for cultivating GEVR crops (Fuchs and Gonsalves, 2007; Prins *et al.*, 2008). Other issues, including those described below, are still considered to be major constraints for testing and deregulating GEVR crops in developing countries.

A. Intellectual property rights

During the green revolution in the 1960s, advances in research toward production of high yielding cultivars (hybrids as well as composites) by industrialized nations were accessible to public sector research institutes in developing countries in Asia, Africa, and Latin America. Agricultural research related to GE crops is now primarily in the domain of multinational companies. These companies hold patents for a vast majority of the currently used technologies, including CP-mediated resistance against plant viruses. The significance of patents to research on GE crops is discussed by Adcock (2007) and Kesan (2007). Developing countries do not have the resources to meet the costs associated with the generation and deregulation of GE crops (Byerlee and Fischer, 2002). Thus, cooperation between multinational companies and public sector research organizations in developing countries is paramount for the production of GE crops of interest to developing countries. Possibilities for accessing the technology that benefit marginal farmers, without paying royalties or limited to nominal costs on humanitarian grounds, remains a workable option (www.ISAAA.org/39-2009) although technologies generated domestically in developing countries are likely to be favored over those originating from multinational companies (Davidson, 2008).

B. Biosafety regulations

Current biosafety regulatory systems were designed largely to deal with the development of a new technology and meet the initial needs of industrialized nations. They are primarily intended to prevent products, considered to be unsafe, from entering the markets. Various criteria have been used to identify threats to public health and environmental safety. Regulations should be formulated in a transparent and consistent manner, taking into consideration the concerns of developers, distributors, farmers, and consumers of the products. Policies adopted in the USA differ from those adopted in the EU (Karplus and Deng, 2008; Pehu and Ragasa, 2007). Biosafety regulations are in place in several countries in Asia, Africa, and Latin America. They have evolved differently across the countries. History, public opinion, economic, and trade incentives and considerations arising from the geographical location have influenced the formulation of these regulations (Karplus and Deng, 2008). South Africa is one of the few countries increasingly aware of the critical need to implement a sensible regulatory framework and a national biotechnology policy. Steps recently taken by the Malawi government in East Africa to implement a national biotechnology policy can be cited as an example of sincere efforts by a developing country toward commercialization of GE crops (www.ISAAA.org/39-2009).

C. Cost of generation of GE crops and enforcement of biosafety regulations

The cost of GE crop production and enforcement of biosafety regulations are some of the most important limiting factors to field test, as well as to deregulate GE crops in developing countries. For example, India spends nearly US$ 500 million per annum on agricultural research from which only US$ 50 million are allocated to research in biotechnology. In contrast the Monsanto company invests over US$ 490 million in biotechnological research (Rao and Dev, 2009). The contribution of developing countries to global biotechnology research spending is less than 9% (Pingali and Raney, 2005). Many of the seed companies in developing countries prefer to purchase licenses to grow and utilize GE crops rather than invest in generating GEVR crops (Evenson, 2004; Pingali and Traxler, 2002). Transfer of the resistance to locally adapted crop cultivars through conventional breeding is often preferred since the necessary skills are available in local public institutions and private sector to achieve this goal (Evenson, 2004). The People's Republic of China is unique as a nation in that it invests heavily in biotechnologies and has become the foremost among developing countries to adopt GE crops (Karplus and Deng, 2008).

Multinational companies are averse to develop technologies for orphan crops such as sorghum, pearl millet, and groundnut/peanut for marginal farmers in developing countries in view of the expenditure involved. A comparative study of costs to enforce biosafety regulations in India and the People's Republic of China showed that the cost for multinational companies in India is high relative to the costs for public sector research institutes to introduce GE crops with the same trait (Pray *et al.*, 2006). The costs estimated to deregulate Bt cotton in India for one transformation event is nearly US$ one million as opposed to US$ 90,000 in China.

D. Campaigns by non-governmental activists

Several organizations operating from different countries oppose the development of GEVR crops (Murray, 2008). Their main argument is that GE crops threaten irreparable damage to human health and the environment (Stone, 2002). One of them, Greenpeace, was chiefly responsible for preventing the deregulation of GE PRSV-resistant papaya in Thailand, despite the fact that the crop performed well in field trials and its release was fully supported scientifically (Davidson, 2008). Some of the countries in Asia have been important centers of anti-GE activities. Apart from running sustained campaigns against Bt cotton (Karplus and Deng, 2008; Kuruganti, 2009; Shiva *et al.*, 1999), Bt brinjal (Centre for Sustainable Agriculture, 2006) and other crops, activists also resort to vandalizing the trial plots. Additionally news media gave wide publicity to local visits by activists. A good example is the coverage given by a widely circulated daily newspaper to the public address in Hyderabad, India, in January 2009 by Mr. Jeffrey Smith, Executive Director, Institute for Responsible Technology, on the health risks stemming from consuming foods from GE crops (Hindu, 2009). Added to this a moratorium on field testing of GE plants is likely to turn out to be a stumbling block for realizing the full potential of GE crops in increasing food production in India.

V. FUTURE PROSPECTS FOR DEREGULATING GENETICALLY ENGINEERED VIRUS-RESISTANT CROPS IN DEVELOPING COUNTRIES

Most scientists, international organizations such as the Consultative Group for International Agricultural Research (CGIAR, http://www.cgiar.org/

impact/agribiotech.html; Okusu, 2009), FAO and World Health Organization (www.who.int/foodsafety/biotech/meetings/animals_2007) and some multinational companies advocate strengthening of biotechnology research to hasten the production and deregulation of GE crops in developing countries. On the contrary some environmentalists, non-governmental activists, and governments, with few exceptions, have rejected the field testing, as well as the commercialization of GE crops. Their contention has been that GE crops have not proven to be safe and multinational companies control their commercialization. As a consequence, marginal farmers are unable to afford or benefit from growing these crops. Among the factors listed in Section IV, biosafety regulations that are scientifically sound and acceptable to governmental agencies should receive high priority. Procedures to assess the safety of GE crops are expensive. By imposing costs at each and every stage of the development of a novel crop, incentives to produce these crops by a variety of developers and distributors will diminish. Ruling governments should support biosafety regulations, which are based on sound scientific experimental evidence. Lessons from the cultivation of GEVR crops in developing countries and extensive safety data (Fuchs and Gonsalves, 2007) could be used as relevant information to facilitate risk-assessment studies in developing countries. Research on various ways to commercialize a GE crop as economically as possible should receive high priority. Partnership with multinational companies and academic research institutions is a viable option to access patented technologies at an affordable cost. It is absolutely essential to increase funding for production and deregulation of GE crops, including the creation of infrastructure for implementing biosafety protocols (Huang *et al.*, 2008) to enable farmers to obtain the major share of the gains from GE crops developed by multinational companies (Karplus and Deng, 2008; Rao and Dev, 2009; World Bank report, 2007). The international agricultural research supported by the CGIAR played a key role in the development and dissemination of technologies in developing countries that lead to the green revolution (Evenson and Gollin, 2003). Funding for these institutes has gradually been reduced and they no longer play a vital role in tapping the full potential of gene revolution technologies (Dalrymple, 2008). There is an urgent need to strengthen this network to harness the GEVR technology for the needs of subsistence farmers in developing countries, especially since technologies generated inhouse are favored relative to those accessed from other countries (Davidson, 2008).

The anti-GE activists have played a significant role in preventing the deregulation of GE crops in developing countries. Legislation to prevent vandalism by these groups should attract appropriate legal action by governmental agencies. Educating extension agencies in various aspects of the technology to produce GE crops will help interaction with farmers to achieve productive results. Many farmers in developing countries are concerned that GE crops grown for export are likely to be rejected in the majority of European countries, reducing the value and marketability of agricultural outputs. The solution of course lies in the acceptance by these nations of products from GE crops.

A number of GEVR crops are in the pipeline for deregulation (FAO, 2008b). Their release will facilitate affordable and secure supply of much needed food in developing countries. Last but not the least it is prudent to promote independent socio-economic research on issues that impact poverty alleviation and inequality (Smale *et al.*, 2009). To this extent, the release of GEVR papaya in Hawaii is an enlightening case study on the adoption of the technology and its positive socio-economic impact (Gonsalves *et al.*, 2004, 2007).

ACKNOWLEDGMENTS

We are thankful to Dr. Bright O. Agindotan of University of Illinois, Champaign-Urbana, for providing us with reprints of numerous publications. We are grateful to Dr. Basavaprabhu Patil of the Donald Danforth Plant Science Center, St. Louis for providing the latest review articles on cassava mosaic geminiviruses and several reprints concerning important tropical viruses.

REFERENCES

Abalo, G., Tongoona, P., Derera, J., and Edema, R. (2009). A comparative analysis of conventional and marker-assisted selection methods in breeding Maize streak virus resistance in maize. *Crop. Sci.* **49:**509–520.

Abhary, M., Patil, B. L., and Fauquet, C. M. (2007). Molecular biodiversity, taxonomy, and nomenclature of tomato yellow leaf curl viruses. *In* "Tomato Yellow Leaf Curl Virus Disease" (H. Czosnek, ed.), pp. 85–118. Springer, New York.

Adcock, M. (2007). Intellectual property, genetically modified crops and bioethics. *Biotechnol. J.* **2:**1088–1092.

Akad, F., Dotan, N., and Czosnek, H. (2004). Trapping of Tomato yellow leaf curl virus (TYLCV) and other plant viruses with a GroEL homologue from the whitefly *Bemisia tabaci*. *Arch. Virol.* **149:**1481–1497.

Anderson, J. M., Palukaitis, P., and Zaitlin, M. (1992). A defective replicase gene induces resistance to cucumber mosaic virus in transgenic tobacco plants. *Proc. Natl. Acad. Sci. U. S. A.* **89:**8759–8763.

Antignus, Y., Vunsh, R., Lachman, O., Pearlsman, M., Maslenin, L., Hananya, U., and Rosner, A. (2004). Truncated Rep gene originated from *Tomato yellow leaf curl virus-Israel* confers strain-specific resistance in transgenic tomato. *Ann. Appl. Biol.* **144:**39–44.

Asano, M., Satoh, R., Mochizuki, A., Tsuda, S., Yamanaka, T., Nishiguchi, M., Hirai, K., Meshi, T., Naito, S., and Ishikawa, M. (2005). Tobamovirus resistant tobacco generated by RNA interference directed against host genes. *FEBS Lett.* **579:**4479–4484.

Atiri, G. I., Winter, S., and Alabi, O. J. (2003). Yam. *In* "Virus and Virus-like Diseases of Major Crops in Developing Countries" (G. Loebenstein and G. Thottappilly, eds.), pp. 249–268. Kluwer Academic Publishers, Dordrecht.

Audy, P., Palukaitis, P., Slack, S. A., and Zaitlin, M. (1994). Replicase mediated resistance to potato virus Y in transgenic tobacco plants. *Mol. Plant-Microbe Interact.* **7:**15–22.

Barone, A., and Frusciante, L. (2007). Molecular marker-assisted selection for resistance to pathogens in tomato. *In* "Marker Assisted Selection: Current Status and Future Perspectives in Crops" (E. C. Guimaraes and B. D. Scherf, eds.), pp. 152–164. FAO, Rome.

Baulcombe, D. (1996). Mechanisms of pathogen-derived resistance to viruses in transgenic plants. *Plant Cell* **8:**1833–1844.

Beachy, R. N., Loesch-Fries, S., and Tumer, N. E. (1990). Coat protein-mediated resistance against virus infection. *Annu. Rev. Phytopathol.* **28:**451–474.

Boonrod, K., Galetzka, D., Nagy, P. D., Conrad, U., and Krczal, G. (2004). Single-chain antibodies against a plant viral RNA-dependent RNA polymerase confer virus resistance. *Nat. Biotechnol.* **22:**856–862.

Bosque-Perez, N. A. (2000). Eight decades of *Maize streak virus* research. *Virus Res.* **71:**107–121.

Braun, C. J., and Hemenway, C. L. (1992). Expression of amino-terminal portion or full-length viral replicase genes in transgenic plants confers resistance to potato virus X infection. *Plant Cell* **4:**735–744.

Büchen-Osmond, C. (2007). *The universal virus database Version 4.* Columbia University, New York.

Byerlee, D., and Fischer, K. (2002). Accessing modern science: Policy and institutional options for agricultural biotechnology in developing countries. *World Development* **30:**931–948.

Calvert, L. A., Koganezawa, H., Fargette, D., and Konate, G. (2003). Rice. *In* "Virus and Virus-like Diseases of Major Crops in Developing Countries" (G. Loebenstein and G. Thottappilly, eds.), pp. 269–293. Kluwer Academic Publishers, Dordrecht.

Capote, N., Pérez-Panadés, J., Monzó, C., Carbonell, E., Urbaneja, A., Scorza, R., Ravelonandro, M., and Cambra, M. (2007). Assessment of the diversity and dynamics of *Plum pox virus* and aphid populations in transgenic European plums under Mediterranean conditions. *Trans. Res.* **17:**367–377.

Carr, J. P., Marsh, L. E., Lomonossoff, G. P., Sekiya, M. E., and Zaitlin, M. (1992). Resistance to tobacco mosaic virus induced by the 54-kDa gene sequence requires expression of the 54-kDa protein. *Mol. Plant-Microbe Interact.* **5:**397–404.

Carr, J. P., Gal-On, A., Palukaitis, P., and Zaitlin, M. (1994). Replicase mediated resistance to cucumber mosaic virus in transgenic plants involves suppression of both virus replication in the inoculated leaves and long distance movement. *Virology* **199:**439–447.

Centre for Sustainable Agriculture (2006). Bt Brinjal – A Briefing Paper, Available at www.csa-India.org/downloads/GE/bt_brinjal_briefing_paper.pdf.

Chaddock, J. A., Lord, J. M., Hartley, M. R., and Roberts, L. M. (1994). Pokeweed antiviral protein (PAP) mutations which permit *E. coli* growth do not eliminate catalytic activity towards prokaryotic ribosomes. *Nucleic Acids Res.* **22:**1536–1540.

Chen, B., Jiang, J. H., and Zhou, X. P. (2007). A *TOM1* homologue is required for multiplication of Tobacco mosaic virus in *Nicotiana benthamiana*. *J. Zhejiang Univ. Sci. B* **8:**256–259.

Cipriani, G., Fuentes, S., Bello, V., Salazar, L. F., Ghislain, M., and Zhang, D. P. (2001). Transgene expression of rice cysteine proteinase inhibitor for the development of resistance against sweetpotato feathery mottle virus. *In* CIP Program Report 1999–2000. International Potato Center, Lima, pp. 267–271.

Crowdy, S. H., and Posnette, A. F. (1947). Virus diseases of cacao in West Africa: II Cross-immunity experiments with viruses 1A, 1B and 1C. *Ann. Appl. Biol.* **34:**403–411.

Cuellar, W. J., Kreuze, J. F., Rajamaki, M., Cruzado, K. R., Untiveros, M., and Valkonen, J. P. T. (2009). Elimination of antiviral defense by viral RNase III. *Proc. Natl. Acad. Sci. U. S. A.* **106:**10354–10358.

Czosnek, H., Ghanim, M., and Ghanim, M. (2002). Circulative pathway of begomoviruses in the whitefly vector *Bemisia tabaci* – insights from studies with *Tomato yellow leaf curl virus. Ann. Appl. Biol.* **140**:15–231.

Dahal, G., Ortiz, R., Tenkouano, A., d'A Hughes, J., Thottappilly, G., Vuylsteke, D., and Lockhart, B. E. L. (2000). Relationship between natural occurrence of banana streak badnavirus and symptom expression, relative concentration of viral antigen, and yield characteristics of some micropropagated *Musa* spp. *Plant Pathol.* **49:**68–79.

Dai, S., Wei, X., Alfonso, A. A., Pei, L., Duque, U. G., Zhang, Z., Babb, G. M., and Beachy, R. N. (2008). Transgenic rice plants that overexpress transcription factors RF2a and RF2b are tolerant to rice tungro virus replication and disease. *Proc. Natl. Acad. Sci. U. S. A.* **105:**21012–21016.

Dale, P. J., Clarke, B., and Fontes, E. M. G. (2002). Potential for the environmental impact of transgenic crops. *Nature Biotechnol.* **20:**567–574.

Dalrymple, G. D. (2008). International agricultural research as a global public good: Concepts, the CGIAR experience, and policy issues. *J. Int. Dev.* **20:**347–379.

Dasgupta, I., Malathi, V. G., and Mukherjee, S. K. (2003). Genetic engineering for virus resistance. *Curr. Sci.* **84:**341–354.

Davidson, S. N. (2008). Forbidden fruit: Transgenic papaya in Thailand. *Plant Physiol.* **147:**487–493.

Deom, C. M., Oliver, M. J., and Beachy, R. N. (1987). The 30-Kilodalton gene product of Tobacco mosaic virus potentiates virus movement. *Science* **237:**389–394.

Diaz-Pendón, J. A., Truniger, V., Nieto, C., García-Mas, J., Bendahmane, A., and Aranda, M. (2004). Advances in understanding recessive resistance to plant viruses. *Mol. Plant Pathol.* **5:**223–233.

Dupre, P., Henry, M., Posadas, G., Pellegrineschi, A., Trottet, M., and Jacquot, E. (2002). Genetically engineered wheat for *Barley yellow dwarf virus* resistance. *In* "Barley Yellow Dwarf Disease: Recent Advances and Future Strategies" (M. Henry and A. McNab, eds.), pp. 27–28. CIMMYT, Mexico.

Edelbaum, D., Gorovits, R., Sasaki, S., Ikegami, M., and Czosnek, H. (2009). Expressing a whitefly GroEL protein in *Nicotiana benthamiana* plants confers tolerance to tomato yellow leaf curl virus and cucumber mosaic virus, but not to grapevine virus A or tobacco mosaic virus. *Arch. Virol.* **154:**399–407.

Efron, Y., Kim, S. K., Fajemisin, J. M., Marek, J. H., Tang, C. Y., Dabrowski, Z. T., Rossel, H. W., and Thottappilly, G. (1989). Breeding for resistance to maize streak virus, a multidisciplinary team approach. *Plant Breed.* **103:**1–36.

Ehara, Y., Nakamura, S., Yoshikawa, M., Shirasawa, N., and Taira, H. (1994). Resistance to viruses in transgenic tobacco plants introduced mammalian 2,5-oligoadenylate synthetase cDNA. *Tohoku J. Agri. Res.* **44:**1–6.

Elmer, J. S., Brand, L., Sunter, G., Gardiner, W. E., Bisaro, D. M., and Rogers, S. G. (1988). Genetic analysis of the tomato golden mosaic virus. II. The product of the AL1 coding sequence is required for replication. *Nucleic Acids Res.* **16:**7043–7060.

Evenson, R.E. (2004). GMOs: Prospects for productivity increases in developing countries. *J. Agrl. & Food Industrial Organization* 2, Article 2.

Evenson, R. E., and Gollin, D. (2003). Assessing the impact of the green revolution. *Science* **300:**758–762.

FAO. (2008a). State of Food Insecurity in the World, 2008 FAO. "*Food Security Statistics*". www.fao.org/es/ess/faostat/foodsecurity/index_en.htm

FAO. (2008b). (http://www.fao.org/biotech/inventory_admin/dep/).

Fauquet, C., and Fargette, D. (1990). African cassava mosaic virus; etiology, epidemiology and control. *Plant Dis.* **74:**404–411.

Fauquet, C. M., Briddon, R. W., Brown, J. K., Moriones, E., Stanley, J., Zerbini, M., and Zhou, X. (2008). Geminivirus strain demarcation and nomenclature. *Arch. Virol.* **153:**783–821.

Febres, V. J., Lee, R. F., and Moore, G. A. (2008). Transgenic resistance to *Citrus tristeza virus* in grape fruit. *Plant Cell Rep.* **27:**93–104.

Fermin, G., and Gonsalves, D. (2003). Papaya. *In* "Virus and Virus-like Diseases of Major Crops in Developing Countries" (G. Loebenstein and G. Thottappilly, eds.), pp. 497–518. Kluwer Academic Publishers, Dordrecht.

Fofana, I. B. P., Sangare, A., Ndunguru, J., Kahn, K., and Fauquet, C. M. (2003). Principles for control of virus diseases in developing countries. *In* "Virus and Virus-like Diseases of Major Crops in Developing Countries" (G. Loebenstein and G. Thottappilly, eds.), pp. 31–54. Kluwer Academic Publishers, Dordrecht.

Fomitcheva, V. W., Schubert, J., Saalbach, I., Habekuß, A., Kumlehn, J., and Conrad, U. (2005). Bacterial expression and characterization of a single-chain variable fragment antibody specific to several replicases of plant (+)RNA viruses. *J. Phytopathol.* **153:**633–639.

Fuchs, M. (2008). Plant resistance to viruses: Engineered resistance. *In* "Encyclopedia of Virology" (B. W. J. Mahy and M. H. V. van Regenmortel, eds.), 3rd edn., vol. 4, pp. 156–164. Elsevier, Maryland Heights, MO.

Fuchs, M., and Gonsalves, D. (2007). Safety of virus-resistant transgenic plants two decades after their introduction: Lessons from realistic field risk assessment studies. *Annu. Rev. Phytopathol.* **45:**173–202.

Fuchs, M., Cambra, M., Capote, N., Jelkman, W., Kundu, J., Laval, V., Martelli, G. P., Minafra, A., Petrovic, N., Pfeiffer, P., Pompe-Novak, M., Ravelonandro, M.,

Saldarelli, I., Stussi-Garaud, C., Vigne, E., and Zagrai, I. (2007). Safety assessment of transgenic plums and grapevines expressing viral coat protein genes: New insights into real environmental impact of perennial plants engineered for virus resistance. *J. Plant Pathol.* **89:**5–12.

Goldbach, R., Bucher, E., and Prins, M. (2003). Resistance mechanisms to plant viruses: An overview. *Virus Res.* **92:**207–212.

Golemboski, D. B., Lomonossoff, G. P., and Zaitlin, M. (1990). Plants transformed with a tobacco mosaic virus nonstructural gene sequence are resistant to the virus. *Proc. Natl. Acad. Sci. U. S. A.* **87:**6311–6315.

Gonsalves, D. (1998). Control of Papaya ringspot virus in papaya: A case study. *Annu. Rev. Phytopathol.* **36:**415–437.

Gonsalves, C., Lee, D. R. and Gonsalves, D. (2004). Transgenic virus-resistant papaya: The Hawaiian 'Rainbow' was rapidly adopted by farmers and is of major importance in Hawaii today. APSnet Features. http://www.apsnet.org/online/feature.

Gonsalves, C., Lee, D. R., and Gonsalves, D. (2007). The adoption of genetically modified papaya in Hawaii and its implications for developing countries. *J. Dev. Stud.* **43:**177–191.

Gordon, D. T. and Thottappilly, G. (2003). Maize and sorghum. *In* "Virus and Virus-like Diseases of Major Crops in Developing Countries" (G. Loebenstein and G. Thottappilly, eds.), pp. 295–334. Kluwer Academic Publishers, Dordrecht.

Gutierrez-Campos, R., Torres-Acosta, J. A., Saucedo-Arias, L. J., and Gomez-Lim, M. A. (1999). The use of cysteine proteinase inhibitors to engineer resistance against potyviruses in transgenic tobacco plants. *Nat. Biotechnol.* **17:**1223–1226.

Hamilton, A. J., and Baulcombe, D. C. (1999). A species of small antisense RNA in posttranscriptional gene silencing in plants. *Science* **286:**950–952.

Hampton, R. O., and Thottappilly, G. (2003). Cowpea. *In* "Virus and Virus-like Diseases of Major Crops in Developing Countries" (G. Loebenstein and G. Thottappilly, eds.), pp. 355–376. Kluwer Academic Publishers, Dordrecht.

Harrison, B. D., Zhou, X., Otim-Nape, G. W., Liu, Y., and Robinson, D. (1997). Role of a novel type of double infection in the geminivirus-induced epidemic of severe cassava mosaic in Uganda. *Ann. Appl. Biol.* **131:**137–148.

Hayakawa, T., Zhu, Y., Itoh, K., Kimura, Y., Izawa, T., Shimamoto, K., and Toriyama, S. (1992). Genetically engineered rice resistant to rice stripe virus, an insect-transmitted virus. *Proc. Natl. Acad. Sci. U. S. A.* **89:**9865–9869.

Henry, M. and Adams, M. J. (2003). Other cereals. *In* "Virus and Virus-like Diseases of Major Crops in Developing Countries" (G. Loebenstein and G. Thottappilly, eds.), pp. 337–354. Kluwer Academic Publishers, Dordrecht.

Henry, M., and McNab, A. (2002). Barley yellow dwarf disease. *In* "Recent Advances and Future Strategies", pp. 139. CIMMYT, Mexico.

Hill, J. H. (2003). Soybean. *In* "Virus and Virus-like Diseases of Major Crops in Developing Countries" (G. Loebenstein and G. Thottappilly, eds.), pp. 377–395. Kluwer Academic Publishers, Dordrecht.

Hindu. (2009). Genetically modified foods a health hazard. January 13, page 3.

Hong, Y., Saunders, K., Hartley, M. R., and Stanley, J. (1996). Resistance to geminivirus infection by virus-induced expression of dianthin in transgenic plants. *Virology* **220:**119–127.

Hong, Y. G., Robinson, D. J., and Harrison, B. D. (1993). Nucleotide sequence evidence for the occurrence of three distinct whitefly-transmitted geminiviruses in cassava. *J. Gen. Virol.* **74:**237–244.

Hsu, H. T. (2002). Biological control of plant pathogens. *In* "Encyclopedia of Pest Management" (D. Pimentel, ed.), pp. 68–70. M. Dekker, New York, Basel.

Huang, J., Zhang, D., Yang, J., Rozelle, S., and Kalaitzandonakes, N. (2008). Will the biosafety protocol hinder or protect the developing world: Learning from China's experience. *Food Policy* **33:**1–12.

Hull, R. (2009). "Comparative Plant Virology", 2nd edn. pp. 269–300. Elsevier Academic Press, New York.

ISAAA. (2009) International Service for the Acquisition of Agri-Biotech Applications. http://www.isaaa.org/resources/publications/briefs/39/executivesummary/

James, C. (2009). Global status of commercialized biotech/GM crops: 2008. The first thirteen years, 1996 to 2008.

Jeeva, M. L. (1997). Effect of secondary infection of cassava mosaic disease on growth and tuber yield of cassava. *J. Mycol. Plant Pathol.* **27:**78–80.

Johnson, S. R., Strom, S., and Grillo, K. (2007). Quantification of the impacts on US agriculture of biotechnology-derived crops planted in 2006. National Center for Food and Agriculture policy. http://www.ncfap.org.

Karplus,V. J., and Deng, X. W. (2008). "Agricultural Biotechnology in China – Origin and Prospects", pp. 165. Springer, New York

Kesan, J. (2007). "Agricultural Biotechnology, Intellectual Property Protection: Seeds of Change", pp. 416. CABI, London

Kosova, K., Chrpova, J., and Vaclav, S. (2008). Recent advances in breeding of cereals for resistance to barley yellow dwarf virus- a review. *Czech J. Genet. Plant Breed* **44:**1–10.

Kreuze, J. F., Klein, I. S., Lazaro, M. U., Chuquiyuri, W. J. C., Morgan, G. L., Mejía, P. G. C., Ghislain, M., and Valkonen, J. P. T. (2008). RNA silencing mediated resistance to a crinivirus (Closteroviridae) in cultivated sweetpotato (*Ipomoea batatas*) and development of sweetpotato virus disease following co-infection with a potyvirus. *Mol. Plant Pathol.* **9:**589–598.

Kreuze, J. F., Valkonen, J. P. T., and Ghislain, M. (2009). Genetic engineering. *In* "The Sweetpotato" (G. Loebenstein and G. Thottappilly, eds.), pp. 41–63. Springer, The Netherlands.

Kuruganti, K. (2009). Bt cotton and the myth of enhanced yields. *Economic and Political Weekly* **44:**29–33.

Lapidot, M., and Friedmann, M. (2002). Breeding for resistance to whitefly-transmitted geminiviruses. *Ann. Appl. Biol.* **140:**109–127.

Lapidot, M., Gafny, R., Ding, B., Wolf, S., Lucas, W. J., and Beachy, R. N. (1993). A dysfunctional movement protein of tobacco mosaic virus that partially modifies the plasmodesmata and limits virus spread in transgenic plants. *Plant J.* **4:**959–970.

Lecoq, H. (2003). Cucurbits. *In* "Virus and Virus-like Diseases of Major Crops in Developing Countries" (G. Loebenstein and G. Thottappilly, eds.), pp. 665–688. Kluwer Academic Publishers, Dordrecht.

Lee, R. F., and Bar-Joseph, M. (2003). Graft-transmissible diseases of citrus. Characteristics of the pathogens, economic impact, and management strategies. *In* "Virus and Virus-like Diseases of Major Crops in Developing Countries" (G. Loebenstein and G. Thottappilly, eds.), pp. 607–639. Kluwer Academic Publishers, Dordrecht.

Lindbo, J. A., and Dougherty, W. G. (1992). Pathogen-derived resistance to a potyvirus: Immune and resistant phenotypes in transgenic tobacco expressing altered forms of a potyvirus coat protein nucleotide sequence. *Mol. Plant-Microbe Interact.* **5:**144–153.

Loebenstein, G., and Carr, J. P. (2006). "Natural Resistance Mechanisms of Plants to Viruses". 532pp. Springer, Dordrecht, The Netherlands.

Loebenstein, G., and Manadilova, A. (2003). Potatoes in the Central Asian Republics. *In* "Virus and Virus-like Diseases of Major Crops in Developing Countries" (G. Loebenstein and G. Thottappilly, eds.), pp. 195–222. Kluwer Academic Publishers, Dordrecht.

Loebenstein, G., and Thottappilly, G. (2003). "Virus and Virus Diseases of Major Crops in Developing Countries". pp. 800. Kluwer Academic Publishers, Dordrecht.

Loebenstein, G., Thottappilly, G., Fuentes, S., and Cochen, J. (2009). Virus and phytoplasma diseases. *In* "The Sweetpotato" (G. Loebenstein and G. Thottappilly, eds.).

Lodge, J. K, Kaniewski, W. K., and Tumer, N. E. (1993). Broad-spectrum virus resistance in transgenic plants expressing pokeweed antiviral protein. *Proc. Natl. Acad. Sci. U. S. A.* **90:**7089–7093.

Lomonossoff, G. P. (1995). Pathogen-derived resistance to plant viruses. *Annu. Rev. Phytopathol.* **33:**323–343.

Lucioli, A., Noris, E., Brunetti, A., Tavazza, R., Valentino, R., Castillo, A. G., Bejarano, E. J., Accotto, G. P., and Tavazza, M. (2003). *Tomato yellow leaf curl Sardinia virus* Rep-Derived resistance to homologous and heterologous geminiviruses occurs by different mechanisms and is overcome if virus-mediated transgene silencing is activated. *J. Virol.* **77:**6785–6798.

Maiti, I. B., Murphy, J. F., Shaw, J. G., and Hunt, A. G. (1993). Plants that express a potyvirus proteinase gene are resistant to virus infection. *Proc. Natl. Acad. Sci. U. S. A.* **90:**6110–6114.

Makkouk, K. M., Kumari, S. G., d'A Hughes, J., Muniyappa, V., and Kulkarni, N.K. (2003). Other legumes. Faba bean, chickpea, lentil, pigeonpea, mungbean, black-gram, lima bean, horsegram, bambara groundnut and winged bean. *In* "Virus and Virus-like Diseases of Major Crops in Developing Countries" (G. Loebenstein and G. Thottappilly, eds.), pp. 447–476. Kluwer Academic Publishers, Dordrecht.

Mao, Y. B., Cai, W. J., Wang, J. W., Hong, G. J., Tao, X. Y., Wang, L. J., Huang, Y. P., and Chen, X. Y. (2007). Silencing a cotton bollworm P450 monooxygenase gene by plant-mediated RNAi impairs larval tolerance of gossypol. *Nat. Biotechnol.* **25:**1307–1313.

Marano, M. R., and Baulcombe, D. (1998). Pathogen-derived resistance targeted against the negative-strand RNA of tobacco mosaic virus: RNA strand-specific gene silencing? *Plant J.* **13:**537–546.

Maule, A. J., Caranta, C., and Boulton, M. I. (2007). Sources of natural resistance to plant viruses: Status and prospects. *Mol. Plant Pathol.* **8:**223–231.

McHale, L., Tan, X., Koehl, P., and Michelmore, R. W. (2006). Plant NBS-LRR proteins: adaptable guards. *Genome Biol.* **7**:212.

Mestre, P., Brigneti, G., Durrent, M. C., and Balucombe, D. C. (2003). Potato virus Y NIa protease activity is not sufficient for elicitation of Ry-mediated disease resistance in potato. *Plant J.* **36:**755–761.

Morales, F. J. (2003). Common bean. *In* "Virus and Virus-like Diseases of Major Crops in Developing Countries" (G. Loebenstein and G. Thottappilly, eds.), pp. 425–445. Kluwer Academic Publishers, Dordrecht.

Moriones, E., and Navas-Castillo, J. (2000). Tomato yellow leaf curl virus, an emerging virus complex causing epidemics worldwide. *Virus Res.* **71:** 123–134.

Murray, I. (2008). The Really Inconvenient Truths. Seven Environmental Catastrophies Liberals Don't Want You to Know about Because They Helped Cause Them. pp. 354. Regnery Publishing Inc., Washington D.C.

New Scientist. (2004). Notes, 7 February 2004:181 pp. 2433.

Niu, Q.-W., Lin, S.-S., Reyes, J. L., Chen, K.-C., Wu, H.-W., Yeh, S.-D., and Chua, N.-H. (2006). Expression of artificial microRNAs in transgenic *Arabidopsis thaliana* confers virus resistance. *Nat. Biotechnol.* **24:**1420–1428.

Odame, H., Kameri-Mbote, P., and Wafula, D. (2001). Innovation and policy process: The case of transgenic sweetpotato in Kenya. www.acts.or.ke/publications/Sweetpotatoandinnovationprocess.pdf.

Ogawa, T., Hori, T., and Ishida, I. (1996). Virus-induced cell death in plants expressing the mammalian 2′,5′ oligoadenylate system. *Nat. Biotechnol.* **14:**1566–1569.

Okada, Y., Saito, A., Nishiguchi, M., Kimura, T., Mori, M., Hanada, K., Sakai, J., Miyazaki, C., Matsuda, Y., and Murata, T. (2001). Virus resistance in sweet potato (*Ipomoea batatas* L. (Lam) expressing the coat protein gene of sweet potato feathery mottle virus. *Theor. Appl. Gen.* **103:**743–751.

Okusu, H. (2009). www.agbioforum.org/v12n1/v12n1a07-okusu

Patil, B. L., and Fauquet, C. M. (2009). Cassava mosaic geminiviruses: Actual knowledge and perspectives. *Mol. Plant Pathol.* **10:**685–701.

Pehu, E., and Ragasa, C. (2007). Agricultural Biotechnology Transgenics in agriculture and their implications for developing countries. Background paper for the World Development Report 2008. (www.worldbank.org).

Pingali, P., and Raney, T. (2005). From the green revolution to the gene revolution: How will the poor fare? ESA Working Paper No. 05-09, November, Food and Agriculture Organization of the United Nations, Rome, Viewed on 30.12.2006 (www.fao.org/es/esa).

Pingali, P., and Traxler, G. (2002). Changing locus of agricultural research: Will the poor benefit from biotechnology and privatization trends. *Food Policy* **27:** 223–238.

Pinto, Y. M., Kok, R. A., and Baulcombe, D. C. (1999). Resistance to rice yellow mottle virus (RYMV) in cultivated African rice varieties containing RYMV transgenes. *Nat. Biotechnol.* **17:**702–707.

Polston, J. E. and Hiebert, E. (2007). Transgenic approaches for the control of tomato yellow leaf curl virus. *In* "Tomato Leaf Curl Virus Disease: Management, Molecular Biology, Breeding for Resistance" (H. Czosnek, ed.), pp. 373–390. Springer, Dordrecht.

Powell-Abel, P., Nelson, R. S., De, B., Hoffmann, N., Rogers, S. G., Fraley, R. T., and Beachy, R. N. (1986). Delay of disease development in transgenic plants that express the tobacco mosaic virus coat protein gene. *Science* **232:**738–743.

Pray, C. E., Ramaswami, B., Huang, J., Hu, R., Bengali, P., and Zhang, H. (2006). Costs and enforcement of biosafety regulations in India and China. *Int. J. Technol. Global.* **2:**137–157.

Prins, M., Laimer, M., Noris, E., Schubert, J., Wassenegger, M., and Tepfer, M. (2008). Strategies for antiviral resistance in transgenic plants. *Mol. Plant Pathol.* **9:**73–83.

Presley, G. J., and George, P. (1999). Banana, breeding, and biotechnology. Commodity advances through banana improvement project research, 1994–1998. World Bank report. Washington D.C. 49 pp.

Qu, J., Ye, J., and Fang, R. (2007). Artificial microRNA-mediated virus resistance in plants. *J. Virol.* **81:**6690–6699.

Rao, N. C., and Dev, M. S. (2009). Biotechnology in Indian Agriculture: Potential, Performance and Concerns. *Academic Foundation,* New Delhi. 198 pp.

Rao, K. V., Rathore, K. S., Hodges, T. K., Fu, X., Stoger, E., Sudhakar, D., Williams, S., Christou, P., Bharathi, M., Bown, D. P., Powell, K. S., Spence, J., Gatehouse, A. M. R., and Gatehouse, J. A. (1998). Expression of snowdrop lectin (GNA) in transgenic rice plants confers resistance to rice brown planthopper. *Plant J.* **15:**469–477.

Razdan, M. K. (2003). *Introduction to Plant Tissue Culture.* 2nd ed. Science Publishers, USA 376pp

Reddy, D. V. R., and Thirumala-Devi, K. (2003). Peanut virus diseases. *In* "Virus and Virus-like Diseases of Major Crops in Developing Countries" (G. Loebenstein and G. Thottappilly, eds.), pp. 397–423. Kluwer Academic Publishers, Dordrecht.

Reddy, D. V. R., Wightman, J. A., Beshear, R. J., Highland, B., Black, M., Sreenivasulu, P., Dwivedi, S. L., Demski, J. W., McDonald, D., Smith Jr., J.W., and Smith, D. H. (1990). Bud necrosis: a disease of groundnut caused by tomato spotted wilt virus. Information Bulletin No. 31. International Crops Research Institute for the Semi-Arid Tropics. Patancheru, A.P. 502 324. pp. 20.

Reddy, D. V. R., Bragard, C., Sreenivasulu, P., and Delfosse, P. (2008). Pecluvirus. *In* "Encyclopedia of Virology" (B. W. J. Mahy and M. H. V. Van Regenmortel, eds.), Vol. 5, pp. 97–105. Elsevier, Maryland Heights, MO.

Roy, A., Savithri, H. S., and Usha, R. (2003). Spices. *In* "Virus and Virus-like Diseases of Major Crops in Developing Countries" (G. Loebenstein and G. Thottappilly, eds.), pp. 773–789. Kluwer Academic Publishers, Dordrecht.

Ryan, M. D., and Flint, M. (1997). Virus-encoded proteinases of the picornavirus super-group. *J. Gen. Virol.* **78:**699–723.

Rybicki, E. P., and Peterson, G. (1999). Plant virus disease problems in the developing world. *Adv. Virus Res.* **53:**127–175.

Safarnejad, M. R., Fischer, R., and Commandeur, U. (2009). Recombinant antibody mediated resistance against *Tomato yellow leaf curl virus* in *Nicotiana benthamiana. Arch. Virol.* **154:**457–467.

Sahijram, L. (2004). Application of biotechnology to *Musa,* a case study. *In* "The Gleanings in Plant Sciences" (N. R. Swami, N. P. Reddy, T. Pullaiah, G. V. S. Murthy and G. P. Rao, eds.), Eastern Book Corporation, New Delhi.

Sano, T., Nagayama, A., Ogawa, T., Ishida, I., and Okada, Y. (1997). Transgenic potato expressing a double-stranded RNA-specific ribonuclease is resistant to potato spindle tuber viroid. *Nat. Biotechnol.* **15:**1290–1294.

Saunders, K., Salim, N., Mali, V. R., Malathi, V. G., Briddon, S. R., Markham, P. G., and Stanley, J. (2002). Characterisation of Sri Lankan cassava mosaic virus and Indian cassava mosaic virus: Evidence for acquisition of a DNA B component by a monopartite begomovirus. *Virology* **293:**63–74.

Shelly, P., Kushwaha, C., Mishra, A., Singh, V., Jain, R., and Varma, A. (2005). Engineering tomato for resistance to tomato leaf curl disease using viral rep gene sequences. *Plant Cell Tissue Organ Cult.* **83:**311–318.

Shepherd, D. N., Mangwende, T., Martin, D. P., Bezuidenhout, M., Kloppers, F. J., Carolissen., C. H., Monjane, A. L., Rybicki, E. P., and Thomson, J. A. (2007). Maize streak virus-resistant transgenic maize: A first for Africa. *Plant Biotechnol. J.* **5:**759–767.

Shiva, V., Emani, A., and Jafri, A. H. (1999). Globalisation and threat to seed security: Case of transgenic seed trials in India. *Econ. Polit. Wkly.* **34:** 601–613.

Smale, M., Zambrano, P., Gruére, G., Falck-Zepeda, J., Matuschke, I., Horna, D., Nagarajan, L., Yerramareddy, I., and Jones, H. (2009). Measuring the economic impacts of transgenic crops in developing agriculture during the first decade: Approaches, findings and future directions. Policy Review #10, pp. 105. International Food Policy Research Institute, Washington DC, USA.

Smith, G. R., and Rott, P. (2003). Sugarcane. *In* "Virus and Virus-like Diseases of Major Crops in Developing Countries" (G. Loebenstein and G. Thottappilly, eds.), pp. 543–565. Kluwer Academic Publishers, Dordrecht.

Sorho, F., Pinel, A., Traore, O., Bersoult, A., Ghesquiere, A., Hebrard, E., Konate, G., Sere, Y., and Fargette, D. (2005). Durability of natural and transgenic resistances in rice to *Rice yellow mottle virus*. *Eur. J. Plant Pathol.* **112:**349–359.

Stone, G. D. (2002). Both sides now: Fallacies in the genetic-modification wars, implications for developing countries and anthropological perspectives. *Curr. Anthropol.* **43:**611–630.

Stone, R. (2008). China plans $3.5 billion GM crops initiative. *Science* **321:**1279.

Sudarshana, M. R., Roy, G., and Falk, B. W. (2006). Methods for engineering resistance to plant viruses. *In* "Methods in Molecular Biology" (P. C. Ronald, ed.), Vol. 354, pp. 183–195. Academic Press, New York..

Tavladoraki, P., Benvenuto, E., Trinca, S., Demartinis, D., Cattaneao, A., and Galeffi, P. (1993). Transgenic plants expressing a functional single chain Fv antibody are specifically protected from virus attack. *Nature* **366:**469–472.

Taylor, D. R. (1989). Resistance of upland rice varieties to pale yellow mottle virus disease in Sierra Leone. *Int. Rice Res. Newslett.* **14:**11.

Tenllado, F., Garcia-Luque, I., Serra, M. T., and Diaz-Ruiz, J. R. (1995). Nicotiana benthamiana plants transformed with the 54-kDa region of the pepper mild mottle tobamovirus replicase gene exhibit two types of resistance responses against viral infection. *Virology* **211:**170–183.

Tepfer, M. (2002). Risk-assessment of virus-resistant transgenic plants. *Annu. Rev. Phytopathol.* **40:**467–491.

Thomas, J. E., Geering, A. D. W., Dahal, G., Lockhart, B. E. L., and Thottappilly, G. (2003). Banana and plantain. *In* "Virus and Virus-like Diseases of Major Crops in Developing Countries" (G. Loebenstein and G. Thottappilly, eds.), pp. 477–496. Kluwer Academic Publishers, Dordrecht.

Thottappilly, G., Bosque-Perez, N. A., and Rossel, H. W. (1993). Viruses and virus diseases of maize in tropical Africa. *Plant Pathol.* **42:**494–509.

Thottappilly, G., Thresh, J. M., Calvert, L. A., and Winter, S. (2003). Cassava. *In* "Virus and Virus-like Diseases of Major Crops in Developing Countries" (G. Loebenstein and G. Thottappilly, eds.), pp. 107–165. Kluwer Academic Publishers, Dordrecht.

Thresh, J.M., and Fargette, D. (2001). Virus diseases of tropical crops. *In* "Encyclopedia of Life Sciences", pp. 1–9. Nature Publishing Group (www.els.net).

Thresh, J. M., Otim-Nape, G. W., Legg, J. P., and Fargette, D. (1997). African cassava mosaic virus disease: The magnitude of the problem. *Afr. J. Root Tuber Crops* **2:**13–19.

Tricoli, D. M., Carney, K. J., Russell, P. F., McMaster, J. R., Groff, D. W., Hadden, K. C., Himmel, P. T., Hubbard, J. P., Boeshore, M. L., and Quemada, H. D. (1995). Field evaluation of transgenic squash containing single or multiple virus coat protein gene constructs for resistance to Cucumber mosaic virus, Watermelon mosaic virus 2, and Zucchini yellow mosaic virus. *Biotechnology,* **13:** 1458–1465.

Truve, E., Aaspollu, A., Honkanen, J., Puska, R., Mehto, M., Hassi, A., Terri, T. H., Kelve, M., Seppanen, P., and Saarma, M. (1993). Transgenic potato plants expressing mammalian 2′,5′-oligoadenylate synthetase are protected from potato virus X infection under field conditions. *Biotechnology* **11:**1048–1052.

Tsao, T. T. T. (2008). Towards the development of transgenic banana bunchy top virus (BBTV)-resistant banana Plants: Interference with replication. Ph.D. thesis submitted to the Queensland University of Technology, Brisbane, Australia, pp. 219.

Tumer, N. E., Hwang, D. J., and Bonness, M. (1997). C-terminal deletion mutant of pokeweed antiviral protein inhibits viral infection but does not depurinate host ribosomes. *Proc. Natl. Acad. Sci. U. S. A.* **94:**3866–3871.

Urcuqui-Inchima, S., Haenni, A. L., and Bernardi, F. (2001). Potyvirus protein: A wealth of functions. *Virus Res.* **74:**157–175.

Vanderschuren, H., Akbergenov, R., Pooggin, M. M., Hohn, T., Gruissem, W., and Zhang, P. (2007). Transgenic cassava resistance to *African cassava mosaic virus* is enhanced by viral DNA-A bidirectional promoter-derived siRNAs. *Plant Mol. Biol.* **64:**549–557.

van der Vlugt, R. A., Ruiter, R. K., and Goldbach, R. (1992). Evidence for sense RNA-mediated protection to PVY^N in tobacco plants transformed with the viral coat protein cistron. *Plant Mol. Biol.* **20:**631–639.

Wang, M. B., Abbott, D. C., and Waterhouse, P. M. (2000). A single copy of a virus derived transgene encoding hairpin RNA gives immunity to barley yellow dwarf virus. *Mol. Plant Pathol.* **1:**347–356.

Watanabe, Y., Ogawa, T., Takahashi, H., Ishida, I., Takeuchi, Y., Yamamoto, M., and Okada, Y. (1995). Resistance against multiple plant viruses in plants mediated by a double stranded-RNA specific ribonuclease. *FEBS Lett.* **372:** 165–168.

Wilson, T. M. A. (1993). Strategies to protect crop plants against viruses: Pathogen-derived resistance blossoms. *Proc. Natl. Acad. Sci. U. S. A.* **90:**3134–3141.

Whitham, S., Dinesh-Kumar, S. P., Choi, D., Hehl, R., Corr, C., and Baker, B. (1994). The product of the tobacco mosaic virus resistance gene N: Similarity to Toll and the interleukin-1 receptor. *Cell* **78:**1101–1115.

Whitham, S., McCormick, S., and Baker, B. (1996). The *N* gene of tobacco confers resistance to tobacco mosaic virus in transgenic tomato. *Proc. Natl. Acad. Sci. U. S. A.* **93:**8776–8781.

Wintermantel, W. M., and Zaitlin, M. (2000). Transgene translatability increases effectiveness of replicase-mediated resistance to cucumber mosaic virus. *J. Gen. Virol.* **81:**587–595.

World Bank, (2007). *World Development Report 2008: Agriculture for Development* The World Bank, Washington, DC.

Xie, Z., Johansen, L. K., Gustafson, A. M., Kasschau, K. D., Lellis, A. D., Zilberman, D., Jacobsen, S. E., and Carrington, J. C. (2004). Genetic and functional diversification of small RNA pathways in plants. *PLoS Biol.* **2:**E104.

Yamanaka, T., Imai, T., Satoh, R., Kawashima, A., Takahashi, M., Tomita, K., Kubota, K., Meshi, T., Naito, S., and Ishikawa, M. (2002). Complete inhibition of tobamovirus multiplication by simultaneous mutations in two homologous host genes. *J. Virol.* **76:**2491–2497.

Yang, Y., Sherwood, T. A., Patte, C. P., Hiebert, E., and Polston, J. E. (2004). Use of *Tomato yellow leaf curl virus* (TYLCV) Rep gene sequences to engineer TYLCV resistance in tomato. *Phytopathology* **94:**490–496.

Yeh, S. D. and Gonsalves, D. (1994). Practices and perspectives of control of papaya ring spot virus by cross protection. *In* "Advances in Disease Vector Research" (K. F. Harris, ed.), pp. 237–257. Springer-Verlag, New York.

Zagrai, I., Capote, N., Ravelonandro, M., Ravelonandro, M., Cambra, M., Zagrai, L., and Scorza, R. (2008). Plum pox virus silencing of C5 transgenic plums is stable under challenge inoculation with heterologous viruses. *J. Plant Pathol.* **90:**63–71.

Zhang, P., Vanderschuren, H., Fuetterer, J., and Gruissem, W. (2005). Resistance to cassava mosaic virus disease in transgenic cassava expressing antisense RNAs targeting virus replication genes. *Plant Biotechnol. J.* **3:**385–397.

Zilberman, D., Ameden, H., Graff, G., and Qaim, M. (2004). Agricultural biotechnology: Productivity, biodiversity, and intellectual property rights. *J. Agrl. Food Ind. Organ.* 2, Article 3.

INDEX

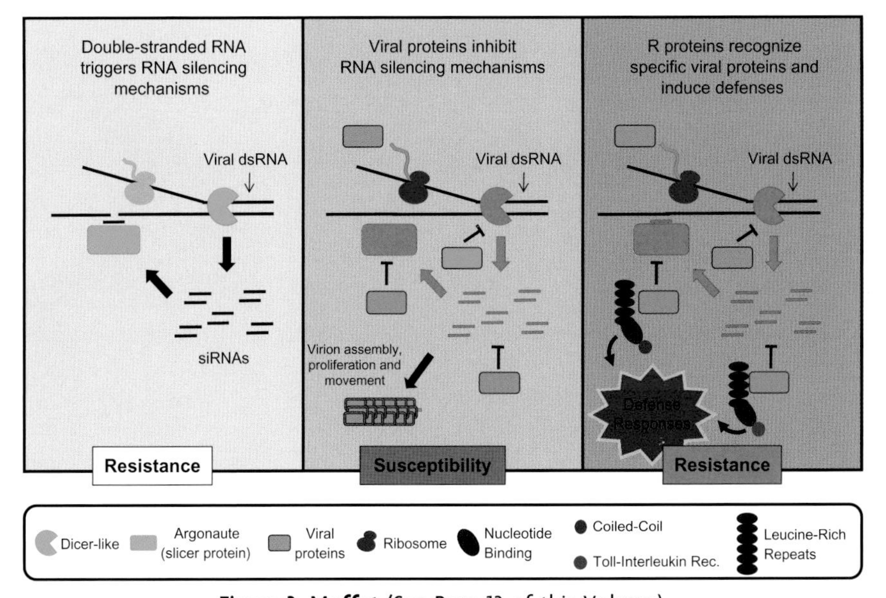

Figure 3, Moffet (See Page 13 of this Volume)

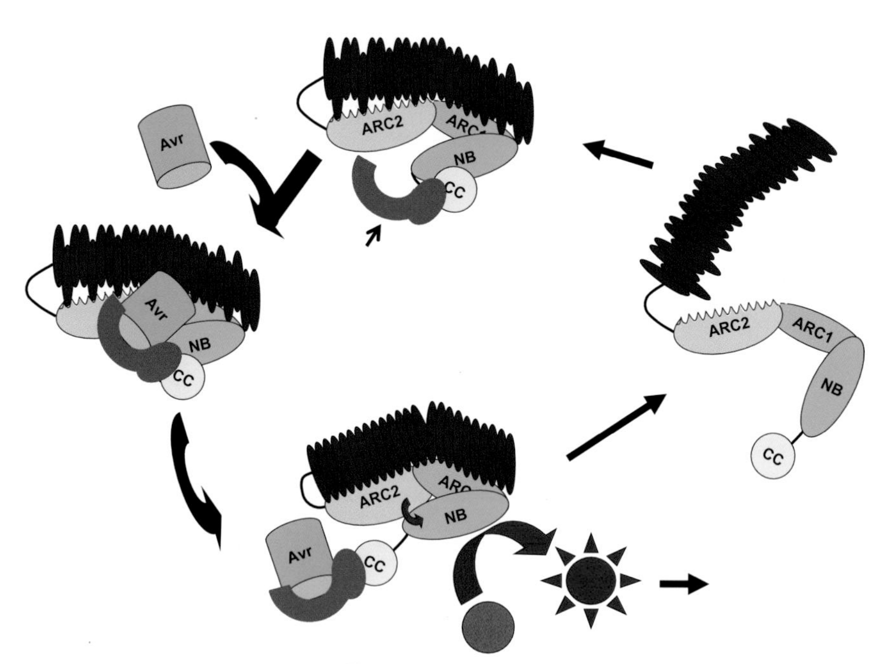

Figure 4, Moffet (See Page 19 of this Volume)

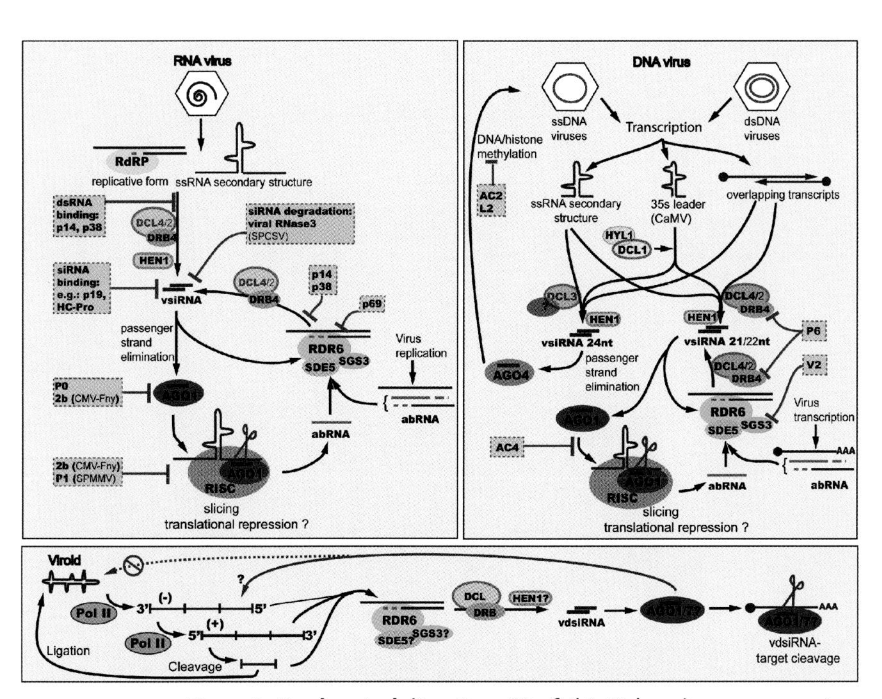

Figure 1, Csorba *et al.* (See Page 39 of this Volume)

(a)

```
Melon     MVVEDSMKATSAEDLSNSIANQNPRGRGGDEDEELEEGEIVGDDD --LDSSNLSAS-LVH 57
Lettuce   MKSEE -QKLIDVNKHRGVRSD-------GEEDEQLEEGEIVGGDADTLSSSSSSRPGTAI 52
Pea       MVVEETPKSIITDDQITTNPN -----RVIEDDNNLEEGEILDED----DSSATSKP-VVH 50
Tomato    MAAAEMERTMSFDAAEKLKAAD---GGGGEVDDELEEGEIVEES----NDTASYLGKEIT 53
Pepper    MATAEMEKTTTFDEAEKVKLN ------ANEADDEVEEGEIVEET----DDTTSYLSKEIA 50
Barley    MAEDTETRPASAGAEER -------------------EEGEIADDG----DGSAAAAAGRVS 38
Wheat     MAEDTETRPASAGAEE R-------------------EEGEIADDG----DGSSAAAAGRIT 38
          *       :                          *****         ..:

Melon     QPHPLEHSWTFWFDNPSAKSKQATWGASIRPIYTFSTVEEFWSVYNNIHHPSKLAMRADL 117
Lettuce   AQHPLEHSWTFWFDTPSAKSKQVAWGSSMRPIY TFSSVEEFWSLYNNIHRPSKLAQGADF 112
Pea       QPHLLENSWTFWFDTPAAKSKQAAWGSSMRPIYTFSTVEEFWSIYNNIHHPGKLAVGADF 110
Tomato    VKHPLEHSWTFWFDNPTTKSRQTAWGSSLRNVYTFSTVENFWGAYNNIHHPSKLIMGADF 113
Pepper    TKHPLEHSWTFWFDNPVAKSKQAAWGSSLRNVYTFSTVEDFWGAYN NIHHPSKLVVGADL 110
Barley    -AHPLENAWTFWFDNPQGKSRAVAWGSTIHPIHTFSTVEDFWSLYNNIHHPSKLNVGADF 97
Wheat     -AHPLENAWTFWFDNPQGKSRQVAWGSTIHPIHTFSTVEDFWGLYNNIHNPSKLNVGADF 97
            * **::******.*   **:  .:**:::: ::***:**:**. *****.*.**   **:

Melon     YCFKHKIEPKWEDPVCANGGKWTVNFPRGKSDNGWLYTLLAMIGEQFDCGDEICGAVVNV 177
Lettuce   YCFKNKIEPKWEDPVCANGGKWTMTFTKAKSDTCWLYTLLAMIGEQFDHGDDICGAVVNV 172
Pea       YCFKHKIEPKWEDPICANGGKWTANYPKGKSDTSWLYTLLAMIGEQFDHGDEICGAVVNV 170
Tomato    HCFKHKIEPKWEDPVCANGGTWKMSFSKGKSDTSWLYTLLAMIGHQFDHGDEICGAVVSV 173
Pepper    HCFKHKIEPKWEDPVCANGGTWKMSFSKGKSDTSWLYTLLAMIGHQFDHEDEICGAVVSV 170
Barley    HCFKDKIEPKWEDPICANGGKWTISCGKGKSDTFWLHTLLALIGEQFDFGDEICGAVVSV 157
Wheat     HCFKNKIEP KWEDPICANGGKWTISCGRGKSDTFWLHTLLAMIGEQFDFGDEICGAVVSV 157
          :***.*********:*****.*. .   :.***. **:****:**.***  *:******.*

Melon     RSGQDKISIWTKNASNEAAQASIGKQWKEFLDYNESIGFIFHDDAKKFDRHAKNKYMV 235
Lettuce   RARQEKIALWTKNAANESAQLSIGKQW KEFIDYNDTIGFIFHEDAKTLDRSAKNKYTV 230
Pea       RGRAEKISIWTKNASNEAAQVSIGKQWKEFLDYNETMGFIFHDDARKLDRNAKNKYVV 228
Tomato    RAKGEKIALWTKNAANETAQVSIGKQWKQFLDYSDSVGFIFHDDAKRLDRNAKNRYTV 231
Pepper    RGKGEKISLWTKNAANETAQVSIGKQWKQFLDYSDSVGFIFHDDAKRLDRNAKNRYTV 228
Barley    RKNQERVAIWTKNAANETAQISIGKQWKEFLDYKDSIGFVVHEDAKRSDKGAKNRYTV 215
Wheat     RQKQERVAIWTKNAANEAAQISIGKQWKEFLDYKDSIGFIVHEDAKRSDKGPKNRYTV 215
          *    ::::*****:**:* * *******:*:**.:::**:.*:**:  *: .**:* *
```

(b)

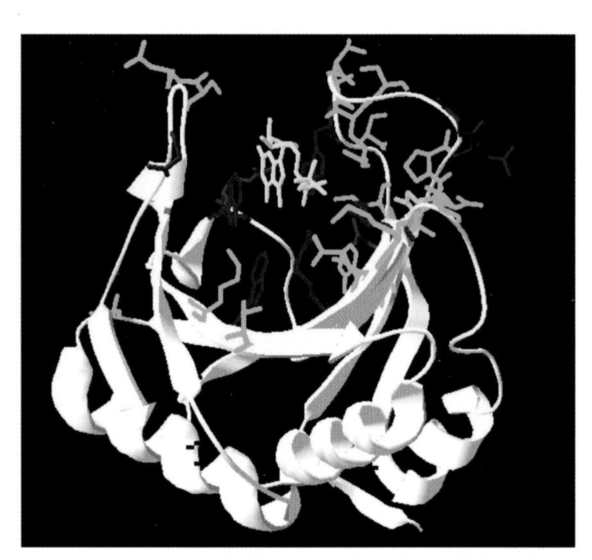

Figure 1, Truniger and Aranda (See Page 137 of this Volume)